Theoremus

Lito Perez Cruz

Theoremus

A Student's Guide to Mathematical Proofs

 Springer

Lito Perez Cruz
Melbourne, VIC, Australia

ISBN 978-3-030-68374-0   ISBN 978-3-030-68375-7 (eBook)
https://doi.org/10.1007/978-3-030-68375-7

This Springer imprint is published by the registered company Springer Nature Switzerland AG
The registered company address is: Gewerbestrasse 11, 6330 Cham, Switzerland

To Lyne, for allowing me to write this

Preface

Today in the teaching of mathematically oriented subjects, students are taught the computational techniques and methods rather than the process of constructing proofs. Unfortunately, when these students are asked to prove a proposition, they are at their wits end. Groping in the dark, they fall all over the place having no clue as to how to proceed. Some attempt to prove but they employ questionable proof methods that make their proofs implausible.

Yet the computational algorithms students are made to learn come from proved theorems. Thus if there is no appreciation of how theorems are proved, mathematics becomes a magical art fogged in mystery. The students may work through the process but they have no idea why the steps they follow work and are valid.

What does it mean then when you are asked to prove a theorem? Essentially, one is asked to write out a proof by argumentation. A proof is a written product, thus writing therefore, is essentially the work involved in the task of proving.

After finishing an undergraduate degree in mathematics, I went back to university to take a degree in the humanities. Not being a good essay writer, I asked one of my professors what was the secret of writing essays? He said, reading! He said the more you read up on your topic, the more you can write something about it. This wisdom rings true too in the art of proving. The more you read proofs, the more you gain necessary knowledge and skills to write proofs too. The secret of writing out proofs is in reading or studying how theorems have been proved.

Why come up with another textbook teaching students how to prove? This book was written for several reasons. Firstly and simply, for students' convenience. There are many thick Discrete Mathematics textbooks that cover proof writing as mainly one of the topics in their extensive reference works. Then there are also textbooks solely designed for helping students how to prove but they too are quite voluminous. An example would be *How To Prove It* [1], which is laudable in many respects. In comparison to these existing works, I have designed this book to be handy enough for students to bring along, read and easily use. My aim is for the student to refer to it when attacking theorem-proving exercises on any mathematical topic. Secondly, this book treats the principles of deduction in greater depth and thus was designed to offer students the opportunity to understand why a particular proof works and why others don't. Furthermore, this work introduces students to Fitch— style proof diagrams which are visually appealing and are easy to follow and understand. Hence, it basically provides students a light introduction to logic.

Additionally, this book wants to provide the student with a minimum reading investment in the art of theorem proof writing. It wishes to assist any student taking mathematical classes in the task of proving mathematical statements. In this work, students will find various examples of proofs and mathematical propositions drawn from number and set theories. The readers do not need to be familiar with such theories. What is given more importance in this book is the style used in proving mathematical statements rather than on the need to have mastery of the subject from which the statements were drawn. Therefore, this book should be useful for all students, whether majoring in mathematics or not. It is for any student taking any quantitative STEM subject requiring the student to provide proofs. Indeed, it is useful for students enrolled in physics, biology, computer science, engineering, economics, banking, finance and the list goes on. Linguistics, did you say? That is included, too!

We have divided the book into two parts, namely, Part I and Part II.

In Part I, we quickly introduce the reader to the basic notion of doing proofs. We swiftly cover the topic in a quick rule of thumb way. We obviously introduce ideas in trickles of logical tools for achieving a proving task. Here we lay down definitions and ground ourselves first with the fundamentals. The objective is that after Part I, the student should be familiar with possible approaches which can be used to confront a proving problem. It is intended that after going through Part I, the reader can now lay it down and apply what was learned to any mathematics subject at hand.

Part II is optional reading. However, it is also a vehicle for the reader to practice what has been learned from Part I. The reader will notice that the basic concepts in Part I are reviewed once again but are given a deeper treatment in Part II. In Part II, we will give examples of how the lessons in Part I are applied. We will also give a method of determining if a statement logically follows from existing known statements, i.e., if it is sound and legally derivable from a set of statements. We do this by first treating mathematically the most fundamental view of logic, which is Propositional Logic. Next we move on to the more expressive type of logic, the one we use in most mathematics, called First-Order Logic or Predicate Logic. The last chapter of Part II should be an interesting reading because there, we explain how theorems in mathematics are discovered and brought to life.

It is my hope that you find this book useful and beneficial.

Melbourne, VIC, Australia   Lito Perez Cruz

Reference

D. J. Velleman, *How To Prove It*. Cambridge University Press, 2006.

Acknowledgements

I received so much help with writing this book on mathematical proofs.

Firstly, my thanks go to my students when I was teaching a theoretical computer science subject at my former university, Monash University, a few years ago. They inspired me to write this book. Without them realizing it, they gave me the ideas found in this work. Thanks should be given too to my university's library where I found reference works relevant to this topic. I recommend them as additional reading to the reader.

I wish to thank Mr. David Pleacher for allowing me to refer to his website at http://www.pleacher.com. The quotations provided comic relief I needed in approaching such a serious subject. I hope the reader takes the humor in a good way.

Several academic colleagues like Dr. James Smith, who gave helpful corrections and Dr. Jim Otto, for useful suggestions deserve thanks as well.

Then there is Colin McAllister, my Internet friend who gave me lots of editorial comments, to him I give my hearty thanks.

Last, but not the least, I thank Ms. Helen Desmond, my editor at Springer who guided me through the process of authoring at Springer Nature. Her patience is so much appreciated.

In the end, I claim responsibility for any errors left remaining in this work and I will be grateful if the reader points them out to me for correction.

Please email them to me at mailto: lpc@sleekersoft.com.

Contents

Acronyms[1]

BC	Before Christ
FOL	First-Order Logic
IH	Inductive Hypothesis
MI	Mathematical Induction
PL	Propositional Logic
RCC	Reduction Ad Absurdum
STEM	Science, Technology, Engineering, Mathematics

---

[1]In general, we define the abbreviations at the place where they are first introduced in the text.

Part I
The Basics

In this part, the readers will look at the basic foundations for proving theorems and other mathematical statements. We first need to give the reader why proofs and proving are a part and parcel of the mathematical discipline, thus, why it is part of the mathematical culture which all mathematicians in general uphold. The readers will also learn that proofs are essentially arguments, and since proofs are arguments, we cannot launch the discussion without first speaking about fallacies. We should be conscious of fallacies so we can avoid falling into them in our own arguments. We devote some time on this then move on to the discussion of types of statements and proofs and so on.

Chapter 1
Introduction

> *"Mathematics is the Queen of the Sciences, and number theory the queen of mathematics."*
>
> —Carl Friedrich Gauss
> &
> "Geeks rule the world"

Abstract In this chapter you will learn that proof is a non-negotiable aspect of the mathematical discipline and no progress is made in mathematics without it. We will emphasize here the truth that proof and the art of proving are significant constituents of the mathematical enterprise.

1.1 The World Has Gone Maths

Finally the world is catching up.

In the olden days you studied subjects like Statistics or Decision Support Systems and back then, you knew well in your mind that taking subjects such as these seemed like a futile exercise. These subjects, you thought, were those that you encounter in and within the academic arena but never relevant somewhere else, and certainly not something you will come across ever again in your adult career life. Well, to our shock, that is not true anymore. Today the commercial world has finally realized there are some use to those dreaded Statistics subjects after all. Operations Research is no longer confined to military use; lots of companies especially in the logistics or freight industry use it often. You might have heard of job titles like Data Analyst, Data Scientist and Financial Quant. Now there are even Marketing Quants! We all know now that these positions have something to do with Mathematics and Statistics.

There is no doubt about it, Mathematics is invading many disciplines and right now even Biology and Psychology are becoming more mathematical. So whoever you may be, if you are obtaining a degree in the sciences, health, finance or economics right now, you will most likely encounter subjects with more mathematical content than you would have ever expected, subjects that may even require you to prove theorems. To illustrate, pick for example a textbook in Econometrics like that of

© Springer Nature Switzerland AG 2021
L. P. Cruz, *Theoremus*,
https://doi.org/10.1007/978-3-030-68375-7_1

Woold [1] and you will find that the author indulges in manual theorem proving. Likewise, the author of a book on Finance like that of Wilmott [2] gives some proofs though these are shared much later in the book. Thus where mathematics is found, it is the case that proofs are found also, by default. So, it is demonstrably a useful skill to know how to prove theorems, even if you are not a math major!

Why do mathematicians prove? Why do they do this sort of thing? Why is this important in mathematics? Let's discuss this.

1.2   The Culture and Tradition of Proofs

Firstly, what is a mathematical proof anyway? Briefly stated, a mathematical proof is an argument composed of a series of assertions, whose aim is to convince the truthfulness of a mathematical statement. When something is true, then it can be used for human advantage. A proof begins from an assumed premise and proceeds by deduction until it reaches finally to the concluding assertion. For brevity, when we say "proof", we mean mathematical proof or argument.

Why do mathematicians have this culture of proving?

In order to find out, we have to look into the nature and history of mathematics. Firstly, it is important to remember that mathematics, as a discipline, deals not only with numbers, but you would have realized after elementary school, it speaks about objects and structures. These objects and structures in general can be abstract. Numbers are just one of those objects it speaks of. It speaks also about the relationship of these objects and structures. The assertions made on these objects and structures must be true and they can be falsified by demonstrating counterexamples. Of all the disciplines in life we can study, mathematics is the only one that has a dogmatic understanding of objective truth. Whereas the other branches of learning in the sciences proceed inductively, mathematics travels on deductively. Whereas the other branches of scientific learning proceed inductively, mathematics on the other hand advances deductively. While we will not be delving into the philosophy of truth in this book, it is nevertheless important for us to be mindful that from its early history, written mathematics has always been concerned with objects and structures that are useful and relatable to our world. Aside from numbers, mathematics deals with entities such as circles, lines, triangles, squares, areas and volumes, among other things—those that we find in nature and in man-made inventions. When we speak about these entities, we invite people to confirm or prove the validity of what we say. Similarly, a proof is a demonstration that what we say about a subject is valid and true.

Secondly, there is the tradition among mathematicians that have valued proof in their craft. Scholars now believe that Indian mathematics pre-dates Greek mathematics as early as 1000 BC. These mathematical ideas spread to the Middle-East, China and Europe. When it comes to proof, archeological documents point to the Western culture, mainly from ancient Greece where first mathematical proofs can be found. The typical character that comes to mind would be the geometrician named

Euclid with his text called the *Elements*. There he treated geometrical objects using the axiomatic (which we will define later) method of proving statements. With new historical documents being discovered lately, scholars of the history of mathematics now believe that the art of proving mathematical statements was also practiced in ancient Eastern and other non-European cultures, like Iran where Islamic mathematicians working on algebra prove theorems we use today. There are also certain ancient Chinese texts that contain what might be considered as mathematical proofs as well [3]. Thus, this proof tradition can be found in all civilizations be they East or West. Proof is an outcome of the nature of things in mathematics.

Whether we like it or not, the tradition of proving mathematical statements is deeply entrenched in the discipline of mathematics. The tradition goes back thousands of years, like 2,500 years and this tradition is hard to break. It is older than us and it will likely go on long after we are gone. Perhaps it will take various forms in the future, like the use of computers to help us in the proving tasks, but mathematical scholars believe the spirit of proof will survive and move on the same way it did in the past.

1.3 Proofs Sharpen Thinking Skills

Having finished high school mathematics would have convinced the reader that mathematics is hard work and needs lots of intellectual investment and energy. Solutions can be hard, not always forthcoming, could be lengthy and can involve trial and error until an exercise problem is solved.

It is a hated subject and I do not deny that more so because it demands so much time from the student. If we are honest, this sometimes produces anger on the student's part. However, this hardship is repaid well with the benefit that the person exercised by it receives strength and stamina to think through arduous problem-solving situations. Yes, the discipline of critical thinking implied in the study of mathematics is applicable to re-contextualized life situations. It helps in making decisions based on information.

So, I encourage the reader to stick with the materials here, be patient with him/herself and work through endurance with the content presented in this humble book. It will eventually click, the penny will drop and you will run with it. Hopefully, what is written here provides you with directions on how to go about the process of solving mathematical proof problems.

Comment

Remember our parts, Part I and Part II. The next chapters will now deal with the actual core concepts. To give a more detailed account we raise the following plan again.

1. In Part I, we will ground ourselves first with definitions, what do we mean by a theorem, what is a proof with a short dealing with fallacies. After this we are now in the position to talk about the types of theorem statements. After classification, we will now have time to spend some foundational principles of doing proofs and this can never be accomplished without rudimentary understanding of logic and its practice. Only then can we give approaches in using several known methods of proving theorems. Here we will give examples of how they are used. We note that this is a quick grounding work which we aim to supply to the reader some bird's eye view of how to go about proving typical mathematical statement forms.
2. In Part II, we apply what we learn in Part I and give the student a chance to go through the math which is applied to logic. So we apply math to logic or logic to logic itself! This part is more formal and the student will get more opportunity to do proof in action. First, we will apply the skills from the previous part to the study of Propositional Logic (PL). It will give the reader a playground area to exercise their knowledge. The takeaway here is that knowledge of PL principles is enough to prove most mathematical statements. However, a more extensive way to handle lots of mathematics like in Calculus or Analysis is knowledge of First-Order Logic (FOL). In both cases, we will show the reader a basic understanding on the "resolution" method, a process for verifying if a statement can be validly concluded from a set of premises.

References

1. J.M. Wooldridge, *Introductory Econometrics*, 5th ed. (South Western Cengage Learning, 2013)
2. P. Wilmott, *Paul Wilmott on Quantitative Finance* (Wiley, England, UK, 2006)
3. K. Chemla, *The History of Mathematical Proof in Ancient Traditions* (Cambridge University Press, 2012).

Chapter 2
Theorems and Proofs

"A mathematician's reputation rests on the number of bad proofs he has given."

—Abram S. Besicovitch

Abstract There are different classifications of mathematical statements like propositions, claims, lemmas, corollaries and theorems. We discuss them in this chapter along with the ideal properties of proofs, about them being valid, rigorous and sound.

2.1 What Are Theorems?

Firstly, what is a theorem? We define a *theorem* as a propositional statement which is considered to be true within the domain of a mathematical system. Yet, that is the question, what is our warrant to say that such and such a statement is true? Theorems advance our body of knowledge in relation to the mathematical system under our study and hopefully they are applicable to some situations in the real world where we move about and roam. Theorems are meant to be applied. They are only of use if they are true because it is dangerous to believe and work on something which is false. It won't work and thus won't be interesting. Thus we are only interested in true theorems and we must be convinced of the their truthfulness. This is where proofs come in.

So theorems are propositions or statements that are deemed to be true. You should have encountered them in your high school geometry subject or algebra. However, there are other terms used to describe mathematical propositions that are true and here are some:

1. Lemma—is a proposition that is used to assist the proof writer in proving a more weightier and more significant theorem. You can think of this as a theorem of smaller impact.

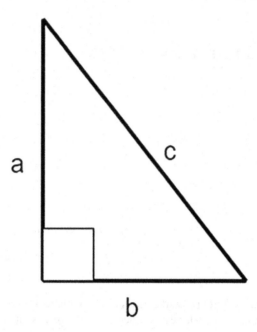

Fig. 2.1 The Pythagorean Theorem

2. Corollary—is a proposition that takes only a few lines of proof arguments to prove. Some authors also use the word "proposition" itself to name this kind of statement.

The choice of naming a proposition is really subjective and can be a matter of taste. The name "theorem" is usually reserved for propositions that have a big impact or revelation but what is "big" is debatable. So the way propositions are named can be a philosophical exercise, so much so some of the mathematicians opt for the term simply as *propositions* instead of theorems. Nevertheless, the above names and the way we described them here can be considered the standard way true statements are named. Remember you see these propositions in the context of a work, manuscript, textbook, monograph, etc. So they flow from the way the mathematician or the way the prover has organized his/her material.

Example

We will now provide a couple of examples of theorems.

The first on our list is the Pythagorean Theorem found below.

Theorem 2.1.1 (Pythagorean Theorem) *Given a right triangle as shown in Fig. 2.1 then*

$$c^2 = a^2 + b^2 \tag{2.1}$$

or

$$c = \sqrt{a^2 + b^2} \tag{2.2}$$

Recall what we said about theorems being useful? The Pythagorean Theorem is a powerful theorem that has seen thousands of applications both in other branches of mathematics itself like Analysis (Calculus) and outside of it. It penetrates almost all branches of Physics. It has been used especially in branches of Engineering like in surveying (geodetic), navigation (maritime) and construction (civil and structural). It is worthy of being called a Theorem and amazingly it was discovered 2,500 years ago.

The next one is not so familiar to us. It is taken from the branch of mathematics called Number Theory. This branch of mathematics is very influential in cryptography so it is useful in the area of digital security. However it is less famous and is not very revelatory.

> **Theorem 2.1.2** (Euclid's 1st Theorem) *Let p be a prime number and let a and b be numbers. If p divides a × b perfectly, then a or b (or both) is divisible by p perfectly.*

Thinking about it some more it is quite trivial and even obvious.

The above is just an example of where the company of mathematicians differs in naming propositions because some call this statement Euclid's Lemma, not Euclid's 1st Theorem.

Thus, some mathematicians today avoid calling such statements "theorems" they call it with another word, like "claims" or "propositions" as we mentioned a few pages ago. One thing is still certain, whatever label they use, the statement still has to be true and thus has to be demonstrated to be so.

In what follows we will take a step back to discuss first arguments which are the foundations of proof. This will ground us better so that the subsequent chapters become easier to absorb.

2.2 What Is an Argument?

We use a set of statements to enjoin people to take upon themselves the reasonability of our position or thesis. We give the reasons why we hold such a statement to be true. When we are doing that we are forming an *argument*. In an argument, we give warrants of why our position on a matter should be accepted by the rest. Here is a classic example we have seen many times before which is the claim *Socrates is mortal* can be argued below:

All men are mortal
Socrates is a man.
_____
Socrates is mortal   ∴

Fig. 2.2 Socrates

We say the above argument is *valid* because it follows the rules of reasoning with respect to the form of the way the argument is set up. The premises support the conclusion, or the conclusion follows from the premises. Yet not all valid arguments are adequate to convince would-be inquirers of the reasonability of one's position or claim. The argument must be *sound* as well. Meaning, that all of the premises are true in fact. The above argument, therefore, is valid and sound because all the premises correspond with the facts. No one indeed lives forever, and Socrates according to witnesses was a human being who was observed and witnesses attest that he indeed died. The story says he drank poison. See Fig. 2.2[1]

Here is an example:

All scientists are never biased
Mr. Bill Nye is a scientist.

Therefore, Mr. Bill Nye is never biased   ∴

The above argument is valid but it is unsound. It will fail to convince hearers. The premise is not always true. Is it true that scientists are never biased? Are scientists exempted from human flaws, are they above opinionated positions? Further, there are those who express a different opinion about Mr. Nye being a scientist, as well.

2.3   What Is a Proof?

We talked about arguments because properly speaking mathematical proofs are arguments too, of a more formal kind. A proof is an argument, therefore, it is composed of a series of assertions, whose aim is to convince the truthfulness of a mathematical statement. Like the reasoning we showed above known as a *syllogism*, it starts from an assumed premise and proceeds by deduction until it finally reaches the concluding assertion. We cannot over emphasize enough the aim—it is to convince cogently the solidness of the reasoning process. Convince who? Convince the one reading the theorem of its validity or convince the mathematical community, or the scientific public, or whoever might profit from its use.

Proofs in mathematics differ from other arguments that can be found in other disciplines like politics, education, philosophy, religion, etc. For example, one will find it very hard to inject emotionally charged words into a proof, there are no "politically correct" speech that must be observed unlike in other arguments outside mathematics. There are no instances where this sort of opportunity can come in when one is doing proofs. There are not "anti-this" or "anti-that" in proofs. We won't see someone who is appalled with the Pythagorean Theorem and is an adamant rejector of it. This is where proofs differ from other arguments.

2.3.1   Fallacious Proofs

We have seen good proof must be valid, meaning it does not violate the rules of reasoning. Also, it should be sound, meaning it is true for every argument for which it speaks.

Before we study further what is a good proof, we should first examine examples of bad arguments and so bad proofs. You must put your critical thinking cap when you are writing proofs and be familiar with fallacious argumentation.

Let us consider, for example, the sixteenth-century belief that *All swans are white*. How did people prove this statement? They reasoned that no one has ever seen any swan that is not white, so all swans must be white. Until, in 1697 Willem de Vlamigh saw some black swans in Western Australia. Before then everyone in the world thought that all swans come in white. The argument being used to prove this was based on experience. Since people have not seen any blue or gray or red swans, etc., then all swans must be white, that was the only kind of swan people saw. This is fallacious, simply because we have not seen unicorns that does not mean unicorns do not exist. The reason we may not believe in unicorns is because of its reputation for being a mythical creature but factually, experience is not always a good guide to use in argumentation.

Today that claim is now false, we now know that there are black swans. Let us see why the above is now a fallacious claim. It is equivalent to saying *There are no non-white swans*. This statement is a "universal existential negative". This claim requires a physical examination to establish its truthfulness. If we do not see non-white swans, who knows we could have been looking at the wrong place. So a fallacy is a logical argument that appears convincing but in the end, it violates either validity or soundness requirements. A fallacious argument contains bad premise that at first glance might sound correct but with deeper analysis prove to be faulty. Most of the time, people do not commit fallacies with dishonest motives, it normally stems from weak critical thinking process.

Proofs and fallacies are quite serious subjects so let us take a break and we wander off to Mr. Pleacher's web page for funny jokes on proofs to help us understand what we are talking about—http://www.pleacher.com/mp/mhumor/proof.html.

I repeat them below so you do not miss out on the laughs. In one way or another I have seen a few of them expressed somewhat similarly by my old professors too. The overused one is the reason—"this case is trivial" and no detailed proof was offered and we moved on to the next one.

If the proof of a theorem is not immediately apparent, it may be because you are trying the wrong approach(sarcasm alert on ;-). Below are some effective methods of proof that may aim you in the right direction.[a]

1. *Proof by Obviousness*: "The proof is so clear that it need not be mentioned."
2. *Proof by General Agreement*: "All in Favor? ...."
3. *Proof by Imagination*: "Well, we'll pretend its true."
4. *Proof by Convenience*: "It would be very nice if it were true, so ...."
5. *Proof by Necessity*: "It had better be true or the whole structure of mathematics would crumble to the ground."
6. *Proof by Plausibility*: "It sounds good so it must be true."
7. *Proof by Intimidation*: "Don't be stupid, of course it's true."
8. *Proof by Lack of Sufficient Time*: "Because of the time constraint, I'll leave the proof to you."
9. *Proof by Postponement*: "The proof for this is so long and arduous, so it is given in the appendix."
10. *Proof by Accident*: "Hey, what have we here?"
11. *Proof by Insignificance*: "Who really cares anyway?"
12. *Proof by Mumbo-Jumbo*: $\forall \epsilon > 0$, $\exists$ a corresponding $\delta > 0$ s.t. $f(x) - L < \epsilon$ whenever $x - a < \delta$.
13. *Proof by Profanity*: (example omitted)

14. *Proof by Definition*: "We'll define it to be true."
15. *Proof by Tautology*: "It's true because it's true."
16. *Proof by Plagiarism*: "As we see on page 238 ...."
17. *Proof by Lost Reference*: "I know I saw this somewhere ...."
18. *Proof by Calculus*: "This proof requires calculus, so we'll skip it."
19. *Proof by Terror*: When intimidation fails ....
20. *Proof by Lack of Interest*: "Does anyone really want to see this?"
21. *Proof by Illegibility*: "⊙ ß ϒ ϖ."
22. *Proof by Logic*: "If it is on the problem sheet, then it must be true."
23. *Proof by Majority Rule*: Only to be used if General Agreement is impossible.
24. *Proof by Clever Variable Choice*: "Let A be the number such that this proof works."
25. *Proof by Tessellation*: "This proof is just the same as the last."
26. *Proof by Divine Word*: "And the Lord said, 'Let it be true,' and it came to pass."
27. *Proof by Stubbornness*: "I don't care what you say! It is true!"
28. *Proof by Simplification*: "This proof reduces to the statement, $1 + 1 = 2$."
29. *Proof by Hasty Generalization*: "Well, it works for 17, so it works for all reals."
30. *Proof by Deception*: "Now everyone turn their backs ...."
31. *Proof by Supplication*: "Oh please, let it be true."
32. *Proof by Poor Analogy*: "Well, it's just like ...."
33. *Proof by Avoidance*: Limit of Proof by Postponement as t approaches infinity.
34. *Proof by Design*: "If it's not true in today's math, invent a new system in which it is."
35. *Proof by Intuition*: "I just have this gut feeling ...."
36. *Proof by Authority*: "Well, Bill Gates says it's true, so it must be."
37. *Proof by Vigorous Assertion*: "And I REALLY MEAN THAT!"
38. *Proof by A.F.K.T. Theorem*: "Any Fool Knows That!"
39. *Proof by vigorous handwaving*: Works well in a classroom.
40. *Proof by seduction*: "Convince yourself that this is true!"
41. *Proof by accumulated evidence*: "Long and diligent search has not revealed a counterexample."
42. *Proof By Blah Blah Blah or Proof by Verbosity*: "blah blah blah ... blah blah blah ... blah blah blah ... finally we have shown what is required."
43. *Proof by Divine Intervention*: "Then a miracle occurs ...."

---

[a]This was compiled by David Pleacher, From Dick A. Wood in *The Mathematics Teacher, November 1998*, Steve Phipps and yours truly.

You can find more laughable and more exhaustive lists than the above if you vigorously search the Internet. Seriously, such types of proofs can be seen in student papers. Most common mistakes made by current students are in the use of intuition in their proof arguments. The use of intuition is a good thing since in fact this is normally what is employed by mathematicians before they formalize their understanding of a problem domain. However that is only the start and it is not where it all ends. The problem with it is that what is intuitive to one person is not intuitive to another. This is also the reason why arguing that something is obvious or it is trivial to prove does not impress certain mathematics teachers. Why? Because what you might consider obvious or trivial can be deemed as a cop out from the real task of proving a theorem. When we invoke the argument of obviousness or triviality we should do so under condition that it is obviously obvious and trivial to the proof reader, your teacher or lecturer. Then in some small extent, we have seen proofs by A.F.K.T. (#38 above), hasty generalizations, plausibility, proof by example or analogy, etc. being offered to no avail.

A common trap committed by students anxious about detail is to fall under the proof by verbosity. Students often wrongly assume that simply because they have written a lot of words, they are entitled to a full mark. Wrong! There are proofs that can be quite long indeed, however, when a proof is too long it loses its effectiveness and such proofs eventually lose the reader along the way. The old Chinese proverb "more talk, more mistake" applies to proofs too. The longer the proof, the more possibility of embedding erroneous statements in the proof. Rather than convince, the long winded deliberation can wind up confusing the reader.

Additionally what teachers and professors hate is a proof wherein the student requires them to fill in the details on the student's behalf. It is certainly most annoying when the student assumes that the proof reader undertstands what the student is on about. Students make the mistake in assuming that their professors can peep into their head. A student who makes the teacher guess what the student wanted to say is asking for trouble. This is most irritating and will most likely incur penalties for lost marks.

We must remember deeply the object of the proof is to convince ourselves and the proof evaluator of the truthfulness or the validity of the proposed statement, i.e., the theorem. We should apply self-criticism to our proof and see to it that it succeeds to convince.

2.3.2 Valid and Sound

One reason why proofs sometimes fail to convince is because it is too informal in its arguments. In contrast to weak proofs or unconvincing proofs, there is what is called *rigorous proofs*. Now this concept may be something vague—mathematicians know one when they see one but they do not know how to describe it in words—perhaps they are overtaken by its beauty and elegance. Let me give a quote from renown mathematicians about this and show how this issue is so emotive:

He[the ideal mathematician] rests his faith on rigorous proof; he believes that the difference between a correct proof and an incorrect one is an unmistakable and decisive difference. He can think of no condemnation more damning than to say of a student, "He doesn't even know what a proof is". Yet he is able to give no coherent explanation of what is meant by rigor, or what is required to make a proof rigorous. In his own work, the line between complete and incomplete proof is always somewhat fuzzy, and often controversial.[2]

Having given an impression that this issue is so subjective, nevertheless, I must attempt to give you an operational notion of what it is, otherwise there will be no improvement in our situation. A rigorous proof provides compelling arguments for the theorem in question. So going back to our description of a proof we briefly mentioned at the start of this chapter, we said it is an argument composed of sequences of assertions starting from a premise. For a proof to be valid and eventually to be rigorous, each assertion must be justifiable. What we mean is that we are standing on reasonable grounds for making the assertions we just made. We mean that when we introduce an assertion that assertion proceeds legitimately according to well established and agreed rules of reasoning. We briefly enumerate the characteristics assertions found in a rigorous proof. We must remember that assertions are as explicit as possible. Here are the characteristics of the subsequent assertions after making the first assertion, which is the premise:

1. They may follow from a definition.
2. They may follow from an axiom.
3. They may follow from previous results or previous proven propositions/theorems.
4. They may follow from an application of the rules of logic or deduction.
5. They may follow no other rule apart from the above.

The above needs a bit of elaborating. By *definitions* we mean an accurate description of a concept or notion or a term in the mathematical subject under discussion. By an *axiom* we mean a statement that is obviously true or held or believed to be always true.

Here is an example:
Having any two points on a plane we can always draw a straight line that passes through them.

This is obviously true. It is an axiom found in Plane Geometry.[3]

Looking at the list above, every movement we make, i.e., from one assertion to the next, proceeds on "legal" grounds. We are saying that any assertion we make to proceed down to the conclusion that we want to prove must follow one of the listed rules. We are not permitted to assert a statement that does not fit any of those characteristics.

In mathematics, valid proofs are sound proofs too. Why? Because definitions assert when a concept fits a property, we have the right to call it with its name and so must be true. Axioms are also statements that are obviously and intuitively true. One

---

[2]Phillip J. Davis and Reuben Hersh, *The Ideal Mathematician*, http://users-cs.au.dk/danvy/the-ideal-mathematician.pdf.

[3]We will come back to this statement later.

can observe and experiment on an axiom and test if it is true, like the one we gave about two points and a line. The rules of logic which we will further elaborate later are also intuitively true. Because of these, then everything in a valid proof is true so all statements are sound as well. This the reason that in proofs soundness is not often discussed because by virtue of the process all valid proofs are sound proofs as well.

2.3.3   Notation Plays a Part

By relieving the brain of all unnecessary work, a good notation sets it free to concentrate on more advanced problems.

—Alfred North Whitehead

You would have known by this time that we human beings use notations or symbols to stand for things or statements. You probably own a mobile device, then you probably also have used *emojies*? Because we are dealing with arguments, sometimes it is much quicker to say things using symbols or signs rather than words. Most of the time we right away recognize something what we see rather than text that we read. For example, we have encountered π before. That is much shorter than writing 3.1416 each time we need to speak about that value in our discussion. If proof has always been part of the mathematical culture and tradition, then the use of notations is too. Arguments can tend to be long if we use natural language in our proof and notations shorten the use of natural language for efficiency. Through the use of notational symbols we are able to write little texts but at the same time convey a lot without the necessity of using proliferation of words. Consequently, the use of symbols in general, aid in making a proof cogent. It becomes a natural part of a formal proof. In a way, this is what makes a proof not only rigorous but elegant or "beautiful". Saying very little and yet meaning a lot is a quality in mathematics that makes it attractive to mathematicians. For some, this is the reason why they are mathematicians today.

As we move forward into the chapters of this work, we will encounter a good use of notational symbols. They do aid in understanding what we are talking about, it makes also the mind attentive and analytical.

2.4   Reflections

Reflection 2.4.1. *Go back to the list of fallacious proof techniques found in Sect. 2.3.1 and examine in what ways you have been guilty of these in the past? Which one of them you frequently committed the most and how will you avoid it next time?*

Reflection 2.4.2. *Proof by example is generally fallacious but there is a time wherein proof by example might be appropriate, in what situation do you think you can use it?*

Chapter 3
Types of Theorems

"A mathematician is a device for turning coffee into theorems."
—Paul Erdos

Abstract In what follows, we will categorize the ways mathematicians state theorems. You should have seen some of them in high school but perhaps, the "new maths" might have prevented you from experiencing that. We should not be bothered that we do not know the details of what these examples speak about. What is important to recognize the form of the propositions rather than understanding the special detail they speak about.

3.1 What Are Theorems?

Theorems as we have seen in the previous sections are statements. They are complete sentences so they convey a complete thought. It is best to see them though as *propositions*, but not any generic propositions. They are propositions that make assertions or they claim something about a subject. For example, if I say *the moon is made of cheese*, I am making a claim. The claim can either be true or false based on the normal use of the language. Most importantly in mathematical parlance, we do not make assertions of a figurative nature, like *the show must go on* or *Mr. Schultz kicked the bucket*, etc. Lets dig down deeper into this in the next section.

3.2 Statements and Propositions

The Merriam-Webster Dictionary says that a *sentence* is a group of words that express a statement, question, command or wish. This is a more precise description than saying that it is an expression of a "complete thought". So clearly then not all sentences are statements.

© Springer Nature Switzerland AG 2021
L. P. Cruz, *Theoremus*,
https://doi.org/10.1007/978-3-030-68375-7_3

19

For our case, a statement is a sentence that conveys a fact, a forceful declaration that must be accepted or rejected. We won't argue with philosophers who make a distinction between a **statement, assertion** or **proposition**. We treat them as synonymous with each other and that they are sentences that can be doubted, believed as true or false. So we have examples:

1. It is raining outside.
2. $x = 1$.
3. $n \in \mathbb{N}$.
4. $S \subseteq T$.

The examples below are not assertions/propositions:

1. Wow! You look gorgeous!
2. What is the value of x?
3. We hope that $x < y$.
4. Find the smallest subset of S.
5. How are you feeling today?

What we want to do for now is to get familiar with propositional statement forms, we will find in theorems. Before we study how to prove them, first lets see how they look like by studying the forms in which we may find them.

3.3 If-Then

This follows the form, if X, then Y. Here X and Y are sentences, i.e., statements that contain a complete thought. Sometimes they are stated differently like this:

1. X, only if Y, or
2. Y whenever X.
3. X implies Y.

Examples

> **Theorem 3.3.1** (Notational Form of Euclid's 1st Theorem) *If a and b are integers and p is prime such that $p|ab$, then either $p|a$ or $p|b$.*

The notation "|" above stands for "divides". So $p|ab$ means p divides the product of a and b. This notation is very common, e.g., see [1] or its treatment at Wolfram MathWorld.

> **Theorem 3.3.2** *If n is an odd integer, then n^2 is odd too.*

The compliment of this theorem is also true, i.e., for even numbers.

Theorem 3.3.3 *If n is an even integer, then n^2 is even too.*

3.4 If and only If

This follows the form, X if and only if Y. This is broken down to two If-Then statements, namely, if X then Y, and if Y then X. This also means statement X is equivalent to statement Y. Sometimes authors abbreviate this as *iff*.

Examples

Theorem 3.4.1 ([2]) *Assume we have a, b as integers. Then b is divisible by a and a is divisible by ba if and only if $a = b$ or $a = -b$.*

Below is an interesting example that says to us that X and Y may be non-trivial propositions [3].

Theorem 3.4.2 *Assume R is a reflexive relation on set A. R is an equivalence relation on A iff $(a, b) \in R$ and $(a, c) \in R$ imply $(b, c) \in R$.*

3.5 Equational Statements

This should be familiar especially in the area of applied mathematics oriented subjects. Simply stated, equational statements are statements in the form of a formula involving the equal sign $=$. It states that the formula on the left or the variable on the left is equal to the formula on the right. Note that the symbol that separates the left side from the right side may be other than the $=$ sign and it may be any relational symbol too, i.e., $\lesssim, \geq, \gg, \Vdash$, etc. In this case we can classify this form as *relational* statements.

Examples

Theorem 3.5.1 (Pythagorean Theorem)

$$c^2 = a^2 + b^2$$

Something from combinatorics

Theorem 3.5.2

$$C(n, r) = \frac{n!}{r!(n-r)!}$$

3.6 Quantified, "for all" Statements

Sometimes we encounter phrases like "for all", "for each", "for every" in the proposition. Such phrases invoke the meaning of generality or universality (or to that effect) and are called *quantifiers* because they denote quantity. For example *All humans are mortal*. Such a statement tells us that mortality applies to every individual that is human. In terms of the number affected by mortality, it applies to everyone. Note that in the said statement, we did not see the word "for" in front of the sentence. It is because we are using our natural language to state a fact. If want to be mathematical, we could restate the same as … "for all x, if x is human, then x is mortal". That would be funny if we spoke that way to our friends. Only when we are in a maths environment do we do math-speak.

Examples

Theorem 3.6.1 (Fundamental Theorem of Arithmetic) *For every $n \in \mathbb{N}$ such that $n > 1$, then n can be expressed uniquely as a product of prime numbers.*

The following is actually a corollary which comes from the above theorem.

Theorem 3.6.2 *Let $n \in \mathbb{N}$ such that $n > 1$, then we can express*

$$n = \prod_{i=1}^{k} p_i^{e_i}$$

where every p_i is prime and each $e_i \in \mathbb{N}$.

Note the variation of expression in that our proposition did not start with the preposition "for". The meaning is the same as saying "For all $n \in \mathbb{N}$ ... etc.".

3.7 Quantified, "there exist" Statements

Here we encounter phrases like "there exists", "there is/are a ..." and the like in the proposition. This is a quantifier that limits the description to a portion of the population domain. The phrase "there exists" implies at least one in number. If the proposition instead has "there is a unique ...", then there is exactly one and only one that matches the description stated in the proposition.

Examples

Theorem 3.7.1 (Division Theorem) *For every $i, j \in \mathbb{N}$, there are unique positive integers q, r such that $i = qj + r$ with $r < j$.*

Note that the "there exists" portion is stated in the last portion of the statement. This is the most typical form we will find in mathematical theories. The form *For all ... there exists* is very typical in mathematical theories. This order is crucial. We may naively think that this ordering may not matter but we must be careful. For the case where the "there exists" comes last, making it occur ahead of the statement may not always work.

UnTheorem 3.7.1 *There is a $c \in \mathbb{N}$, such that $y = x + c$ for every $x < y$ that are in $\mathbb{N}$.*

The above statement says there is a fixed c that works for any $x < y$ and that is a problematic statement. At the very least, it requires more time for mental digestion. So as a rule, do not think that the reversing of the order is benign. Below is a statement where the "there exists" comes at the start of the statement and is always true.

Theorem 3.7.2 *There is a y such that for all x, we have xy = x.*

That y that makes the above statement always true is $y = 1$.

Lastly, so that we may be exposed to some variety of propositions we have a negation of "there exists" below.

Theorem 3.7.3 *There is no injective function $f : X \rightarrow 2^X$ for any set X.*

We won't offer a definition of an injective function but only to illustrate a classic example of a non-existence theorem.

3.8   Reflections

Reflection 3.8.1 *Consider the statement "when we combine two countable sets the resulting set is also countable" [4]. Classify this statement according to our theorem types discussed above.*

Reflection 3.8.2 *What do you think happens when you reverse the order of quantifiers in Theorem 3.7.2? Is the statement still true? If so, can we say that if we reverse the* there exist ... for all *form to* for all ..., there exist *form, the outcome is the same?*

References

1. A. Clark, *Elements of Abstract Algebra* (Dover Publications, 1984)
2. V. Shoup, *A Computational Introduction to Number Theory and Algebra* (Cambridge University Press, 2005)
3. J.L. Mott, A. Kandel, T.P. Baker, *Discrete Mathematics for Computer Scientists and Mathematicians* (Prentic Hall, 1986)
4. T.A. Sudkamp, *Languages and Machines* (Addison-Wesley Publishing Company, 1994)

Chapter 4
Logical Foundations of Proof

"Proof is an idol before whom the pure mathematician tortures himself."

—Arthur Stanley Eddington

Abstract In this chapter we will supply you with some practical lessons in logic which you will normally employ in proving theorems. Although experts will consider our discussion light, yet you will be equipped with tools in the process of proving. We take our metaphor from the mechanic who usually could go away and fix most car issues by the simple use of pliers and screw drivers. You can get away using them but it won't be absolutely efficient. We need more tools. For our case we need to be conversant with two useful classical logics:

1. Propositional Logic
2. First-Order Logic/Predicate Logic.

4.1 Propositional Logic

If we are going to write proofs at all, we cannot escape the use of logic. The foundation of sound reasoning or argumentation is the proper use of this important tool of mathematics. If we do not know the fundamental tenets of logic then the arguments in our proofs will be weak and even down right wrong. For this reason, we should also be aware of logical fallacies and see to it that we do not commit them in our proofs. In Sect. 2.3.1 we humorously encountered them.

© Springer Nature Switzerland AG 2021
L. P. Cruz, *Theoremus*,
https://doi.org/10.1007/978-3-030-68375-7_4

4.1.1 Basic Components

So far we have dealt with **simple** assertions like those we saw in Sect. 3.2.

What about statements like "$x > 5$ and $x < 3$", "either $x > 5$ or $x < 3$", "x is not $= y$", or "if $x > 5$ then x is not less than 3"? Observe that these statements are a combination of two statements which are joined by a connectives like "and", "or", "not", "then", etc. We call the form of these propositions **compound** because there are several assertions related to each other.

Let us now resort to the use of symbols or notations to make our discussion verbally economical and efficient. We will do this so that we can understand better the statements we are analyzing. Note that in the proofs mathematicians make, they do not completely resort to symbols, for the proofs they mostly make are still, to some degree, informal. Unless you are studying a subject in mathematical logic, then and only then do you expect to make all your arguments symbolic or formal, meaning according to form. For the sake of better analysis we will temporarily appeal to the use of symbolic logic though eventually we will not be purely symbolical in our math proofs later. Since we can recognize assertions we will now represent them with letters like p, q, r, etc. We will also symbolize connectives: $\land$ for "and" (conjunction), $\lor$ for "or" (disjunction), $\neg$ for "not" (negation) and $\rightarrow$ for "then/imply" (implication). We will also use the left (and right) parenthesis for grouping propositions the same way we use them in algebra, for priority evaluation (discussed later). We have the following examples, on the left is the assertion and on the right is its symbol:

1. $x > 5$: p.
2. $x < 3$: q.
3. $x = y$: r.

So we have the corresponding symbolical representation of the assertions:

1. $x > 5$ and $x < 3$: $p \land q$.
2. either $x > 5$ or $x < 3$: $p \lor q$.
3. x is not $= y$: $\neg r$.
4. if $x > 5$ then x is not < 3: $p \rightarrow \neg q$.

4.1.2 True/False Valuations

We can see that propositions can be technically long or complicated. For example, we could have this $\neg p \lor q \land r \rightarrow s$ so what now? This is where the parentheses come in and make clear which ones come together.

For example in $\neg p \lor q \land r \rightarrow s$ is it like this $\neg p \lor (q \land r \rightarrow s)$ or like this $((\neg p \lor q) \land r) \rightarrow s$? How we group the propositions together is crucial because the overall proposition has a truth value, i.e., the whole proposition is either true or false. In the end that is what we are talking about.

Table 4.1 Truth table for $\wedge$ and $\vee$

p	$\wedge$	q	p	$\vee$	q
T	T	T	T	T	T
T	F	F	T	T	F
F	F	T	F	T	T
F	F	F	F	F	F

Table 4.2 Truth table for $\neg$ and $\rightarrow$

			p	$\rightarrow$	q
$\neg$		q	T	T	T
F	T		T	F	F
T	F		F	T	T
			F	T	F

So overall every proposition, be complex or simple will have a value, that is either true or false. In maths, we of course are interested in propositions that are useful and so their true/false values provide helpful information in judging their applicability to a problem situation. In Table 4.1, we have what is called a *Truth Table*. If we notice, this is the possible assignments of true or false values to each of the propositions. In it we see what is the result of the final value of the compound proposition. It is also called a *valuation*. So in that table, we have the resulting valuation for conjunction and disjunction. So for example, in Table 4.1, when p is true, and q is also true, then the resulting value of the conjunction of the two, i.e., $\wedge$ is true as well. Some consider these as a form of definition, but we can also consider them to be sensible assignments borne out of our common sense experience. Let us take the example of "and". Let us say "It is raining and it is wet outside". We cannot say this statements true if it is not actually raining outside, meaning one of the clauses in that statement is not true, both have to be true for the meaning of "and" to be true. In Table 4.2 we have the valuation for negation and implication.

Ask yourself when is $p \wedge q$ **TRUE**, when is it **FALSE**? Do the same for $\vee$ and $\neg$. In $\vee$, we mean either one of the propositions is true or both of them are true. The more interesting table is the implication or the *conditional* table. Note when $\rightarrow$ is **FALSE**. The implication statement says that when $\rightarrow$ is **TRUE** and the condition stated by p is **TRUE**, then you should bet the farm that q is **TRUE** also. So for example I can say "If $x = 2$, then I'll eat my dog's breakfast". Here our q bears no common sense relationship with our p and yet for our purpose, we will accept without arguments such statements. Such attitude says that $\rightarrow$ is taken to be *material implication/condition*, i.e., without regard to the connection—real/un-real of q with p. Note also that the statements below are **TRUE**:

1. If 3 is even, then 4 is odd.
2. If 3 is even, then 4 is even.

Table 4.3 Truth table of $((p \vee q) \wedge q) \rightarrow \neg p$

$((p$	$\vee$	$q)$	$\wedge$	$q)$	$\rightarrow$	$\neg p$
T	T	T	T	T	F	F
T	T	F	F	F	T	F
F	T	T	T	T	T	T
F	F	F	F	F	T	T

We know that 3 is never even, the above propositions are always **TRUE**, also known as *vacuous truth*. Vacuously true statements follow the form $p \rightarrow q$ of which it is known that p is false. So when such is the case, it does not matter what q might be, the implication *to* will always be true. See the $\rightarrow$ truth table. In maths of course we are only interested in p and q being **TRUE** without being vacuous. Some areas of pure mathematics and logic allow for vacuous truths. Outside of these disciplines or subjects, vacuous truths have no added informational value. For example, if I say "All of my pets are dogs" (when in fact, I got no pets).

Lastly we have the symbol $\leftrightarrow$ for "if and only if". $A \leftrightarrow B$ is **equivalent** to $(A \rightarrow B) \wedge (B \rightarrow A)$. You can work out the truth value for $\leftrightarrow$ based on what we know of $\wedge$ and $\rightarrow$ and you will find that it is **TRUE** if both A and B are **TRUE** or both of them are **FALSE**. Another phrase used in place of "if and only if" is "A is necessary and sufficient condition for B".

From the above we can see that we can break down a complex proposition into simple propositions which are compounded together.

Example

In this example, we look at Table 4.3. Assume that it is the case that $p \vee q$, and that it is q, does this imply that it is $\neg p$? Our first intuition might say "yes", but let us confirm this by the truth table method. The number of rows is calculated by computing 2^n where n is the number of propositions, in this case 2. We can see that the resulting valuation is not always **TRUE**. When the valuation of a compound proposition is **always TRUE**, we call this a *tautology* and is forever true. For example $p \rightarrow p$ is tautologous, or a tautology. However for our case above, it is not always **FALSE** either. When the resulting valuation is **always FALSE**, we call this a *contradiction*. The situation we have in Table 4.3 is called *contingent*. Upon deeper reflection, we know that $p \vee q$ can be true if both p, q are true. So knowing it is q from $p \vee q$ does not imply it is $\neg p$. To see this, look at our 1st row and note the values of p, q and how that row makes $\rightarrow$ **FALSE**. Likewise knowing it is q from $p \vee q$, does not imply it is p either.

Example

Try forming the truth table for $p \rightarrow (q \rightarrow r) \rightarrow [(p \rightarrow q) \rightarrow (p \rightarrow r)]$. It is tedious and quite complex already, isn't it? This requires 8 rows already for 3 component propositions. Truth table technique for determining if a complex proposition is either (in)consistent (i.e., tautologous) or contingent requires a lot of work. Fortunately, there is another and quicker method for this problem, and it is through the construction of what is called a *refutation tree*. We won't cover it here[1] but I just want you to be aware of such a technique that is covered by most logic textbooks.

4.1.3 Logical Equivalences

In the previous sections we covered the equivalent connective ↔. Some call this the *biconditional*. When the biconditional is tautologous, it is called a **logical equivalence**. This means we can substitute the proposition on the left with the one on the right and we designate this situation with the ⟺. Why is this an important topic to cover? Well if you are asked to prove a theorem, it might be easier instead to prove it's logically equivalent expression. In the process of proving the equivalent, you thereby proved the original form of the theorem. This is the reason why we cannot escape the principles of logic in mathematics. The life of the (pure) mathematician is at least made pleasant by such principles. Below are just some useful logical equivalences (with their names) that could just make your task so enjoyable and fun [1]:

Material Implication/Impl.	$P \to Q \Longleftrightarrow \neg P \lor Q.$
Double Negation/D.N.	$\neg\neg P \Longleftrightarrow P.$
Transposition/Trans.	$P \to Q \Longleftrightarrow \neg Q \to \neg P.$
De Morgan/D.M.	$\neg(P \land Q) \Longleftrightarrow \neg P \lor \neg Q.$
De Morgan/D.M.	$\neg(P \lor Q) \Longleftrightarrow \neg P \land \neg Q.$
Exportation/Exp.	$P \to (Q \to R) \Longleftrightarrow P \land Q \to R.$

You notice that the propositional letters we just used were upper case letters. There is a reason for this. The above forms are *schemas*, meaning they stand for complex propositions. So for example, say we are being asked to prove $(s \land t) \to (u \lor w)$. This is in the form of Impl. So we can prove instead $\neg(s \land t) \lor (u \lor w)$ by identifying $P = (s \land t)$, $Q = (u \lor w)$.

In addition, when we convert natural language statements to symbolic forms, we can refer to them as *formulas*. We call them formulas because we are after their form. So we have the conjunctive $p \land q$, disjunctive $p \lor q$, negative $\neg p$ and implicative $p \to q$ forms or formulas. The concept of schemas is very useful in this regard. When we see a formula, no matter how complicated they might be, we can consider them as schemas of which elaborate combinations of formulas can be formed following its form. Thinking of them also as schemas helps us to consider them as templates. So then, I recommend the reader to treat the logical formulas we see here as schemas so that the reader does not get overwhelmed by what seemly complicated formulas we might find in what follows.

---

[1] You can make a request and I will include it in the next edition.

4.1.4   Inference Rules

By *rules of inference* or *rules of deduction*, we mean the rules that must be followed and applied to premises, i.e., propositions or assertions, in order to make new ones. We emphasize the bit about "making new formulas", because it adds or gives us new information not obvious to us before. In other words, we are reasoning on these formulas.

Before talking about the rules of deduction followed in logic, we need to state and agree to the following principles or *Laws of Thought*:

Law of Identity/LOI   Any statement that is true is true.
Law of Non-Contradiction/LNC No statement can be true and false at the same
                                   time.
Law of Excluded Middle/LEM   A statement can only be true or false.

These laws are actually the laws of classical logic. Just be aware that there are also what is known as "Non-classical" logics that do not necessarily abide by the above laws but we won't get in to that. The non-compliance to the above laws makes one's reasoning faulty so we need to assume that when an assertion is made it is either true or false, nothing in between.

Figure 4.1 shows how one starts from the premise, moves by deduction making assertions until we reach the conclusion, the result we want to prove. In a way the proof steps may be called a *derivation* for we are trying to derive from the premise the resulting conclusion, the object of our proof. The format of the derivation we will follow have been inspired by [2] and is called *Fitch Style Natural Deduction*,[2] just so that you know what it is called by logicians. Fitch Style is a diagrammatic and linear way of showing how the argument starts from the top which are the premises and proceeds downwardly in the process. Each move we make which are assertions must follow from the laws of deduction using theorems proved before, definitions or axioms that we agreed as the foundation of the mathematical domain under question.

Each new assertion must be "legal", i.e., must follow the rules of inference/ laws of deduction, of which we will first name the most used and the most important one, *modus ponens*.

$$
\begin{array}{c|l}
1 & P \to Q \\[4pt]
2 & \vdots \\[4pt]
3 & P \\ \hline
4 & Q
\end{array}
$$

This rule says that if we have $P \to Q$ and that P is the case, we can assert Q anywhere in our proof argument. Technically with modus ponens, our logical equivalences and truth tables, we are done already.

---

[2]Invented by Frederic Brenton Fitch (1908–1987), an American logician.

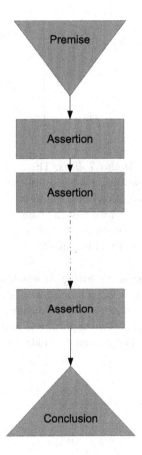

Fig. 4.1 Flow of argument

As an example, let us have the following formulas: $p =$ "you sleep late", $q =$ "you will be tired". So let us form the maxim, "if you sleep late, then you will be tired". Modus Ponens (MP) says then the following:

1	*"if you sleep late"* $\rightarrow$ *"you will be tired"*
2	$\vdots$
3	*"you slept late"*
4	*"you will be tired"*

These elements are sufficient tools for the task of proving but we will list a few more rules which make our life a little more easier [1]:

$$
\begin{array}{c|c}
1 & P \rightarrow Q \\
2 & \vdots \\
3 & \neg Q \\
4 & \neg P \quad\quad MT
\end{array}
\qquad
\begin{array}{c|c}
1 & P \rightarrow Q \\
2 & \vdots \\
3 & Q \rightarrow R \\
4 & \vdots \\
5 & P \rightarrow R \quad\quad HS
\end{array}
$$

In the above, MT stands for *Modus Tollens*. This states that if the implication is accepted and then further down the consequent of that implication is denied, then we are right to conclude the negation of the antecedent. HS stands for *Hypothetical Syllogism* which means that when we have a chain of implications being asserted like the above, then we can assert the head antecedent implies the tail consequent. Some might say this is the property of transitivity. This is analogous to algebra, like $a = b$ and $b = c$, then $a = c$.

Below we have DS for *Dysjunctive Syllogism* which says that if we have accepted a dysjunction and down the tract we got a negation of one of its components, then we have the right to assert the other part of that dysjunction. CD is for *Constructive Dilemma*, It says if we have a set of two implications, and it has been found that either of their antecedents are true, then we can conclude that either of their consequents are true too.

$$
\begin{array}{c|c}
1 & P \vee Q \\
2 & \vdots \\
3 & \neg P \\
4 & Q \quad\quad DS
\end{array}
\qquad
\begin{array}{c|c}
1 & (P \rightarrow Q) \wedge (R \rightarrow S) \\
2 & \vdots \\
3 & P \vee R \\
4 & \vdots \\
5 & Q \vee S \quad\quad CD
\end{array}
$$

Below we have Abs for *Absorbtion*. This is an interesting rule. For it says that if we have an implication along that way that we were able to assert then we have the right to now create a new assertion where the antecedent implies itself in conjunction with its consequent. Simp is for *Simplification* that says if we have a conjunction we have the right to assert either one of the components of that conjunction. It stands to reason that if the conjunction is true, then it can only be that both parts are true and so we can assert either one of them to be true, as well.

$$
\begin{array}{c|l}
1 & P \to Q \\
2 & \vdots \\
3 & \vdots \\
4 & P \to (P \land Q) \quad \textit{Abs}
\end{array}
\qquad
\begin{array}{c|l}
1 & P \land Q \\
2 & \vdots \\
3 & \vdots \\
4 & \vdots \\
5 & Q \qquad \textit{Simp}
\end{array}
$$

Below we have the easy obvious ones. Conj is for *Conjunction* saying that any assertion we make as we go along downwardly, we have the right to form their conjunction. The other is called Add for *Addition*. It says any assertion we make, we can introduce another proposition in dysjunction with that assertion. Thing of it for a moment, if we assert a proposition to be true then whatever we join with it by way of dysjunction is always true as well. Here you might want to review the valuation of ∨ we encountered awhile ago.

$$
\begin{array}{c|l}
1 & P \\
2 & \vdots \\
3 & Q \\
4 & P \land Q \quad \textit{Conj}
\end{array}
\qquad
\begin{array}{c|l}
1 & P \\
2 & \vdots \\
3 & \vdots \\
4 & \vdots \\
5 & P \lor Q \quad \textit{Add}
\end{array}
$$

$$
\begin{array}{c|l}
1 & \vdots \\
2 & P \\
3 & \vdots \\
4 & Q \\
5 & P \to Q \quad \textit{Cond}
\end{array}
$$

This last one above says that if we have at some point in our deduction we deduce P and then at another point further down in the process we arrived at Q, the we have the right to assert it is $P \to Q$. In fact this is what we are somewhat doing when we try to prove theorems of the If-Then form. We start with the condition/antecedent A, make deductions, then arrive at the conclusion/consequent C. At that point we say q.e.d, we are done. Yet, without explicitly mentioning it, we are in fact asserting $A \to C$. We don't formally do this because the theorem to be proved has that as the statement already at the top. The final inference rule we will present is the so called *reductio ad absurdum rule*, meaning "reduced to absurdity". We will label this rule as RAA. Let us say we are trying to prove $P \to Q$ we proceed as follows:

1	P	Given
2	$\neg Q$	Assumption
3	$\vdots$	
4	$R \wedge \neg R$	(contradiction)
5	$\perp$	falsehood
6	Q	(from falsehood you can conclude anything)
7	$P \rightarrow Q$	Cond, 1, 6, ... *RAA*

RAA procedure says we assume P and $\neg Q$ at the same time. In the above if at any point we derive a contradictory statement such as that in 4, then we have a falsehood signified by $\perp$ in 5. The philosophical principle—from a contradiction we can conclude anything takes over. This is also known in Latin as *ex falso quodlibet*. It means from falsehood we can prove anything, so we can also conclude Q. Thus from the principle of Cond, we get $P \rightarrow Q$.

Example

Assume we have $A \wedge B$, let us deduce that $A \wedge B \rightarrow \neg(\neg A \vee \neg B)$. Below is our proof and note the justifications on the right as each step is taken.

Proof

1	$A \wedge B$	Assumption
2	A	1, Simp
3	$\neg\neg A$	2, D.N.
4	B	1, Simp
5	$\neg\neg B$	4, D.N.
6	$\neg\neg A \wedge \neg\neg B$	5, 3, Conj
7	$\neg(\neg A \vee \neg B)$	6, D.M.
8	$A \wedge B \rightarrow \neg(\neg A \vee \neg B)$	1, 7, Cond

□

As we said earlier, mathematicians generally prove theorems informally, or sometimes semi-formally. We will hardly see them line up their proof the same way we just did as shown above. However their arguments follow a linear flow though much prose is often used. Furthermore, the argument steps they make have either explicit or implicit justifications attached to each step or assertion they make. At times they may not invoke the justification verbally because the reader can easily see what the justification that was just used. What is important is that the proof gives the reader a solid and convincing set of arguments. Of course, since mathematicians present their work to their colleagues, the more formal the proof, the more beautiful and elegant it is and often easier to follow and to verify correctness.

4.1.5 Reflections

We have discussed lots of concepts in this section so it is appropriate to stop now a do some exercises.

> **Reflection 4.1.1** *Is* $P \rightarrow (P \rightarrow Q)$ *a tautology? Can you construct a derivation of this?*

> **Reflection 4.1.2** *Can you verify that* $[(P \wedge Q) \rightarrow \neg P] \rightarrow (P \rightarrow Q)$ *is a tautology? Can you construct a derivation of this? Hint: Confirm that* $(\neg P \vee \neg P) \Longleftrightarrow \neg P$. *We will discover that this forms the basis for the* indirect proof *or* proof by contradiction *method, which we will learn later in this book.*

> **Reflection 4.1.3** *Recall the Abs rule we encountered above. Starting from* $P \rightarrow Q$, *use the other rules to prove the assertion* $P \rightarrow (P \wedge Q)$.

4.2 Predicate Logic

Unfortunately, there are some limitations of Propositional Logic (PL). When the sentences are neat and tidy we can always represent them as simple propositions which we can combine to form complex ones. Let us examine what happens when we have the following sentences which we turn into propositions:

A = All men[3] are mortal.
S = Socrates is a man.
M = Therefore, Socrates is mortal.

Our derivation becomes

---

[3]Human beings including women.

$$
\begin{array}{c|c}
1 & A \\
2 & S \\
\hline
3 & M
\end{array}
$$

Suddenly it appears that M just drops in from nowhere, and the deduction looks weird or strange. Hence, PL can be inadequate in handling these types of statements. It is because the statements have an internal structure that gets lost when you simply represent them as simple propositions. S and M look fair enough but A has a word, in this case *all*, that gives more information that must be discerned. Furthermore, when we look at S and M, we note that they follow the subject-predicate pattern. What we mean is that an individual (the subject), such as Socrates, is asserted or "predicated" to have a certain property or belonging to a certain category—like being mortal, being human, being tall, being male/female, etc. (cf. [1, 2]). This is precisely why logicians call the inference we just made above as *categorical syllogism*. The property that certain objects or individuals have is called a *predicate*. Because of these predicates, PL inferences can be weak when handling these more expressive statements, and they are found in every area of mathematics. So we come now to what is called Predicate Logic or what is called First-Order Logic (FOL), the logic that adequately handles whatever mathematical theories we might have.

4.2.1 Basic Components

All the symbols in PL are absorbed in FOL and we shall learn later that they are extended such that FOL becomes a worthy vehicle for dealing with the statements found in mathematical theories. For example, in FOL the grouping provided by parentheses in PL are respected and obeyed, and below we discuss what are those elements in FOL that make it more expressive than PL.

Individuals/Variables

FOL works under an assumed *domain* of discussion. For example in Theorem 3.6.1, the domain is the natural numbers. In Theorem 3.7.3, the domain is about functions. In the sentences we just saw above we talked about humans that was the domain. Then we spoke about an individual in that domain, called Socrates. Logicians call these individuals *constants*. Symbolically lower case letters like a, b, c, d, etc. are used for constants. When we are not naming constants then we are talking about the domain in general and we are speaking about an un-specified individual or *variables* symbolized by x, y, z, u, v, w, etc. These are placeholders for named objects that might occupy them. These constants or variables may be subscripted if we run out of letters.

Predicates/Relations

We represent predicates with upper case letters, like P, Q, R, S, etc. or we may use abbreviations too. So, for example, we can say H for being human, or Mom for being a mother, etc. So if a = Socrates, and we want to represent "Socrates is human", we get $H(a)$. So if j = Jane, and we want to represent "Jane is a mother", we get $Mom(j)$.

What happens now if the predicate has some relational notion? Like "Mary is the wife of Joseph"? Then if W stands for being a wife, then we have the statement symbolized by $W(Mary, Joseph)$. So we can see that a predicate can relate several number of individuals together. Furthermore, the statements we just made such as $H(a)$, $Mom(j)$, $W(Mary, Joseph)$ they give a complete thought, don't they? In other words, they are sentences and they make sense.

Now try this $H(x)$, $Mom(x) \cdot W(x, y)$, what do they convey? Well we are not able to answer this question by just looking at the predicate and the variable. It simply means a thing called x has the property H, Mom, etc. That is vague. So we come now to the discussion about quantifiers.

The Quantifiers

We need to say more about x in $H(x)$. This is where the *quantifiers* come in. Quantifiers quantify the variables and are signaled by words like *all, every, some, at most, at least*, etc. In FOL, we add two connectives, namely, $\forall$ (for all, every, all, each), $\exists$ (there exists, there are, there is, some) in front of the predicate. So for example $\forall x H(x)$ or $\exists x Mom(x)$.

So how do we read these examples? For $\forall x H(x)$, we will read "all x such that x has the property of being an H". Some authors put a dot in between the $x H(x)$ like this $\forall x \cdot H(x)$ so that the reader can add the "such that" phrase to the understanding of the symbol. However, in our case we will follow the tradition of omitting this signal and remember to put in the phrase. So for $\exists x Mom(x)$, we will read "there exists an x such that it is a Mom".

These can be preceded by the negative sign. For example, say we have the statement "not all men believe in God" or "there are no one who believes in God". Then we can represent this by $\neg \forall x B(x)$ where B is the predicate standing for "believe in God". Take note that the exact symbolic translation is not going to be smooth.

So we see that FOL has additional quantifier connectives $\forall$ and $\exists$ on top of PL. Aside from this, we now have predicates (or relations) and we also have functions. These elements are traditionally represented by words starting with capital letters (P, Q, R, S, T, H, Mom, Pop, $Child$, etc.) and lower case letters, respectively, (f, g, h). We of course have as stated above variables and constants too. Lastly we should mention that we have a special two-place predicate called *equal* ($=$).

Example

So let us have an example. Take the statement "I said in my haste, 'All men are liars'". This can be translated as $\forall x(M(x) \to L(x))$. Now this is not the same as $\forall x M(x) \to L(x)$, for here, the $\forall$ controls $M(x)$ only and it does not control $L(x)$.

Because of the grouping provided by the parentheses, the x in $L(x)$ is not necessarily the same x mentioned in $M(x)$. In fact the latter can be re-written like $\forall x M(x) \rightarrow L(y)$ and the effect would have been the same. Take note though that the symbolic translation will not always be smooth.

4.2.2 Categorical Forms

Recall that we can look at predicates as categories. When we are asserting an object x to have a property P, we are saying x belongs to a class having the property P. We can classify our quantifier statements into four forms which have been named by medieval logicians by A, E, I, O. We show them below [2]:

A: All P's are Q's $\simeq \forall x(P(x) \rightarrow Q(x))$
E: Some P's are Q's $\simeq \exists x(P(x) \rightarrow Q(x))$
I: No P's are Q's $\simeq \forall x(P(x) \rightarrow \neg Q(x))$
0: Some P's are not Q's $\simeq \exists x(P(x) \wedge \neg Q(x))$.

The formula on the right of $\simeq$ is the symbolic representation of our natural language expression on the left.

4.2.3 True/False Valuations

In PL, we used truth tables or truth assignments. In FOL, it is a bit more complicated than that. Here we need to say when is an FOL statement *satisfiable*, meaning, do we have a domain of discourse objects spoken about by the statements for which the statement is true? For example, say "Unicorns exist"—$\exists x U(x)$. This statement is true if we can demonstrate there is a being which happens to be a unicorn. If that is the case—the statement is "satisfiable". The object x does not have to have a name but say it does and the name of the individual is "Peggy"—symbolized by g. Then we can test if $U(g)$ is indeed true, that is—is Peggy a unicorn? If so then again $\exists x U(x)$ is satisfiable, thus true. The domain of discourse serves as the context where we can evaluate the truthfulness or falsity of the FOL statements.

So how do we know if $\forall x P(x)$ is true? We can examine all the objects and test if $P(x)$ is true for all of them and if there is one object that cannot be said to have the property P, then $\forall x P(x)$ is false. If all the objects make $P(x)$ true then $\forall x P(x)$ is true. Another way of looking at this is to say that we cannot find an x that makes $P(x)$ false. If that is the case $\forall x P(x)$ is true.

4.2.4 Logical Equivalences

In the same way that we have logical equivalences in PL, we also have them in FOL. Here they are

De Morgan	$\neg \forall x\, P(x) \iff \exists \neg P(x).$
De Morgan	$\neg \exists P(x) \iff \forall x \neg P(x).$
Commutativity	$\forall x \forall y\, P(x, y) \iff \forall y \forall x\, P(x, y).$
Commutativity	$\exists x \exists y\, P(x, y) \iff \exists y \exists x\, P(x, y).$
Distribution	$\forall x (P(x) \wedge Q(x)) \iff \forall x\, P(x) \wedge \forall x\, Q(x).$
Distribution	$\exists x (P(x) \vee Q(x)) \iff \exists x\, P(x) \vee \exists x\, Q(x).$

This means that when we are in the process of deduction, whenever we see the form of the statement in the left side, we can replace it with the one on the right and vice versa.

We need to be careful with Distribution. $\forall x (P(x) \vee Q(x)) \not\iff \forall x\, P(x) \vee \forall x\, Q(x)$. You cannot distribute $\forall$ when the connective is $\vee$. Likewise we cannot distribute $\exists$ if the connective is $\wedge$. We can quickly see where Distribution does not work. For example, let us look at the natural numbers $\mathbb{N}$. Let O be an "integer is odd", and E be an "integer is even". Clearly it is true that $\forall x (O(x) \vee E(x))$, that is all integers are either odd or even. Yet this is not equivalent to $\forall x\, O(x) \vee \forall E(x)$, for the this says either all numbers are odd or all numbers are even. This we know is not true, because we know that some integers are odd and some integers are even.

A helpful tool to better understand categorical statements is the Square of Opposition (SoO), Copi [1] has a treatment of this and we show this in diagram[4] form shown in Fig. 4.2.

How do we read this and how will it help us gain a quick intuition of how these categorical forms relate? In the diagram, the black areas mean there are zero elements in that set. The bright red areas mean there are some elements in these areas.

So, SaP and SeP being contraries mean that they can both be false but cannot both be true. SiP and SoP being subcontraries mean that they can both be true but cannot both be false. SaP and SoP are contradictories mean that if SaP is true then SoP is false and vice versa. Similarly if SeP is true then SiP must be false since both are contradictories and vice versa. Now if SaP is true then by observation, SiP must be true too. The latter is subaltern of SaP, meaning, part of what is asserted is true. Then, this means that of course SoP is false (since SaP and SoP are contradictories) and thus SeP must be false too (since SiP and SeP are contradictories). Assume now we are given that SeP is false. Then SiP must be true but we cannot determine if SaP is true/false neither can we determine SoP. SiP being subcontrary to SoP and vice versa means that the two may be true but both cannot be false at the same time. Though "Some S are P" it does not imply that most definitely that "All S are P" neither would it imply that "Some S are not P". The Venn diagram shown in Fig. 4.3 might be of help in seeing this situation.

---

[4]Credits: Watchduck (a.k.a. Tilman Piesk), Public domain, via Wikimedia Commons.

Reflection 4.2.1 *What happens to the rest if we are given that O is true? if we are given E is true?*

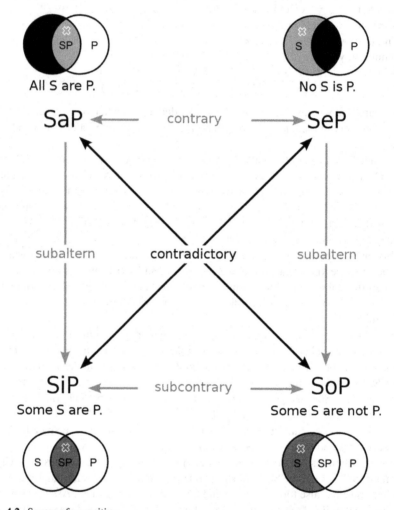

Fig. 4.2 Square of opposition

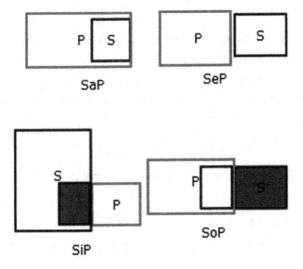

Fig. 4.3 Square of opposition using sets

4.2.5 Inference Rules

For our inference rules we need to add a few more to those we have in Sect. 4.1.4.

$$
\begin{array}{ll}
1 & \forall x\, P(x) \\
2 & \vdots \\
3 & P(a) \qquad \text{UI}
\end{array}
\qquad
\begin{array}{ll}
1 & P(a) \\
2 & \vdots \\
3 & \forall x\, P(x) \qquad \text{UG (Caution!!!)}
\end{array}
$$

$$
\begin{array}{ll}
1 & \exists x\, P(x) \\
2 & \vdots \\
3 & P(a) \qquad \text{EI (Caution!!!)}
\end{array}
\qquad
\begin{array}{ll}
1 & P(a) \\
2 & \vdots \\
3 & \exists x\, P(x) \qquad \text{EG}
\end{array}
$$

We have followed the names given by [1] for the derivation. UI stands for *Universal Instantiation*. It stands to reason that if everyone has the property P, then it must be true for every individual a in the domain of discourse. So, it must be true for a member of the domain, this is what it says

UG stands for *Universal Generalization*. We have a caution in that the individual a should have been chosen arbitrarily. The best way to see this is to keep track on how a has been introduced in the derivation. We need to watch out for this.

EI stands for *Existential Instantiation* and a has not been used before. Lastly EG stands for *Existential Generalization* where a is any individual.

Example

We shall show that the following syllogism is valid. *Everyone who seeks will find.
Cain does not find. ∴ Cain does not seek.*

Proof Let S be for "someone who is a seeker", F for "someone who finds". The
first sentence is $\forall x(S(x) \rightarrow F(x))$. Let c stand for Cain. Then the second premise
becomes $\neg F(c)$. The conclusion to prove is $\neg S(c)$.

1	$\forall x(S(x) \rightarrow F(x))$	Given
2	$\forall x(\neg F(x) \rightarrow \neg S(x))$	1, Trans, cf. Sect. 4.1.3
3	$\neg F(c) \rightarrow \neg S(c)$	UI, 2
4	$\neg F(c)$	Given
5	$\neg S(c)$	Modus Ponens, 3, 4

□

4.3   Fallacy Alert

In the next chapter we will give examples of how to prove mathematical statements
so you will read a number of proofs. Do you recall in Sect. 2.3.1 we covered the
humorous side of fallacious proofs? Actually there is a serious side to that. Logicians
have categorized a swag of fallacious arguments and it will be useful if you study
some of them, to avoid them. We said that mathematicians use arguments in proofs
and as students of mathematics, therefore arguments are what we present. In saying
that assertions must proceed from legal grounds, it also means they must not be
fallacious. The best way is to be self-critical of your own arguments.

So for example, when you make a claim, re-read it and ask your self the question,
could I be guilty of a fallacy here? What is my justification for my claim? Such
self-questioning attitude keeps you safe in the straight and narrow path. Fallacies
are so easy to commit, even experienced mathematicians sometimes commit them.
So put your critical hat on when reading proofs, and most specially, when you are
writing them.

References

1. I. Copi, *Introduction to Logic* (The Macmillan Company, 1969)
2. J. Barwise, J. Echemendy, *Language, Proof and Logic* (Seven Bridges Press, 1999)

Chapter 5
Types of Proof Techniques

"A math lecture without a proof is like a movie without a love scene. This talk has two proofs."

—H. Lenstra

Abstract In this chapter we will use the intellectual equipment we just saw previously to establish proofs for theorems. We said that the secret to writing proofs is the reading of proofs. Therefore, we have plenty of examples to read in this chapter where we classify several types of statements and how to prove them.

5.1 Direct Method

Assume that we are to prove a theorem of the form $P \to Q$. The direct method simply means that we start from premise P, move forward with deductions from it until we land on the conclusion Q. Of course in maths, it is taken for granted that we may employ the previous theorems that have been proved in our argument in the process of proving $P \to Q$. So we will now have examples.

Examples

Theorem 5.1.1 *Let $a, b, c \in \mathbb{N}$, i.e., integers. If a divides b, then a divides bc.*

Proof First we ask, what does it mean to say that "a divides b"? This means that $b/a = k$ for some integer k. This means $b = ak$. Consider now bc. We have $bc = akc$ thus $bc/a = akc/a = kc$, showing that a divides bc as required. $\square$

It is worth mentioning the obvious; one of the ways to move from P to Q is to rely on definitions and theorems that have been proven before. We shall do something like this in the next example below.

© Springer Nature Switzerland AG 2021
L. P. Cruz, *Theoremus*,
https://doi.org/10.1007/978-3-030-68375-7_5

Theorem 5.1.2 *The set of even numbers is denumerable (countably infinite).*

Proof Let $\mathbb{E}$ be the set of even numbers. By definition, $\mathbb{E}$ is countably infinite or denumerable iff the cardinality of $\mathbb{E}$, $|\mathbb{E}| = |\mathbb{N}|$. Consider now the function

$$f : \mathbb{E} \to \mathbb{N}$$

with f being

$$f(e) = e/2.$$

Since e is even then it is divisible by 2, and $f(e) \in \mathbb{N}$. So for each $e \in \mathbb{E}$ there is a $y = f(e) \in \mathbb{E}$. Also let $f(e) = y_1$ and $f(e) = y_2$. $y_1 = f(e) = e/2$ and $y_2 = f(e) = e/2$, thus $y_1 = y_2$, based on this f is a total function. Further if $e_1 \neq e_2$, we have $f(e_1) \neq f(e_2)$, hence f is one-to-one. Since there is a total one-to-one function from $\mathbb{E}$ to $\mathbb{N}$, then $|\mathbb{E}| = |\mathbb{N}|$ which implies $\mathbb{E}$ is denumerable by definition.   □

We can also look at the form of the above theorem as proving an object x satisfies the property Q, i.e., $Q(x)$. We can say that we are moving from property $P(x)$ to showing also that $Q(x)$ is the case.

5.2 Indirect Method

Assume we are asked to prove $P \to Q$. The indirect route assumes $\neg Q$ then do deductions so as to arrive at $\neg P$. Since we know that by Transposition in Sect. 4.1.3, $P \to Q$ is equivalent to $\neg Q \to \neg P$, $(P \to Q \leftrightarrow \neg Q \to \neg P)$ then, it is deemed to have proven the former if we have proven the latter. So one more time …

Theorem 5.2.1 *Let $a, b, c \in \mathbb{N}$, i.e., integers. If a divides b, then a divides bc.*

Proof Assume that a does not divide bc then $bc/a \neq k$, for all k. Then $b/a \neq l$ for any l. This implies that a does not divide b, for if it does then $bc/a = m$ for some m. Then by virtue of Transposition, we have proven the theorem as required.   □

Note that in this proof we are not saying anything about c/a. We are not concerned with that. What we want is to say something about a and b, because they are mentioned in the proposition as the premise whose negative becomes our consequent in this method of proof.

5.3 Proof by Contradiction

The indirect proof has a strong relationship with this type of proof and in fact some authors do not distinguish between proof by contradiction and the indirect method of proof. The main tool for this approach of proving is the RAA rule found in Sect. 4.1.4. Recall how we were able to prove $P \rightarrow Q$ by assuming contrary facts. Proof By Contradiction is one of the most powerful tools at the mathematician's disposal. For if the direct proof is showing to be a difficult path to track, perhaps we can see the light clearly if we follow the RAA method of proof. We offer the following examples.

Theorem 5.3.1 *Let n be an odd integer, then n^2 is odd as well.*

Proof Assume n is odd but n^2 is even (thus not odd). Then $n^2 = 2k$, for some k. Since n is odd the we can express $n = 2m + 1$ for some integer m. Then we have
$$\Rightarrow n^2 = 2k$$
$$\Rightarrow (2m + 1)^2 = 2k$$
$$\Rightarrow 4m^2 + 4m + 1 = 2k$$
$$\Rightarrow 4m^2 + 4m = 2k - 1$$
$$\Rightarrow 4m(m + 1) = 2k - 1$$
$$\Rightarrow 2(2m(m + 1)) = 2k - 1.$$
The left side of the final step, $2(2m(m + 1))$ is always an even number since it is a factor of 2, yet the right side $2k - 1$ will always be odd. So we have arrived at a situation wherein an even number = an odd number, contradiction! $\therefore$, n^2 must be odd also. $\qquad \square$

Theorem 5.3.2 *Assume that $y = x^2 - x - a$. Assume further that a be an odd integer. Then y has no integer roots.*

Proof By definition an integer root of y is an integer x such that $y = 0$. Assume that there exists such an x for y. Therefore we have
$$\Rightarrow x^2 - x - a = 0$$
$$\Rightarrow x^2 - x = a$$
$$\Rightarrow x(x - 1) = a.$$
Consider the left side of the last line. Since x is an integer then it is either odd or even. If x is even, then $x - 1$ is odd and thus $x(x - 1)$ is a product of two numbers, an even number multiplied by an odd number. This product is even because any even number multiplied by an odd number is always even, and so we have an even number $x(x - 1) = a$ which is odd. This is a contradiction. On the other hand, if x is odd, then $x - 1$ is even and again $x(x - 1)$ is an even number which is equal to a which

is odd. Again we have a contradiction. Since in all cases we get a contradiction then it must be that no such integral root exists. The reader can refer to [1].   □

If you observe, proof by contradiction works very well for quantified propositions like the above. Demonstration of existence or non-existence of an object most of the time is best achieved through the use of RAA.

5.4 Left-Right Method

This section pertains to *if and only if* statements. When we see a proposition of the form—$P \leftrightarrow Q$, we need to remind ourselves that this is actually a composition of two propositions, namely, $P \rightarrow Q$ (which we label as the "left" side of the iff statement) and $Q \rightarrow P$ (which we label as the "right" side of the iff statement). If we notice, each of them lends itself to the application proof techniques found in this textbook, starting from the direct method in Sect. 5.1. We can proceed to first prove either side of the iff statement, but normally the left side is proven first and then the right side. The choice which of the sides to prove first is entirely a matter of convenience. We can go first for the side that is easier to prove.

In the following example, we recall the definition of equivalence relation R on a set A. R is said to be an equivalence relation if it is reflexive, symmetric and transitive. We refer the reader to [1] for the details and for the example below. The definition found in the Internet may be useful also.

Theorem 5.4.1 *If R is a reflexive relation, then R is an equivalence relation iff $(a, b) \in R$ and $(a, c) \in R$ imply $(b, c) \in R$.*

Proof
$\Rightarrow$ (Left to Right)
$(a, b) \in R$, then $(b, a) \in R$ since R is an equivalence relation and by the symmetric property. Combine this with $(a, c) \in R$ we get $(b, c) \in R$ by the transitive property of an equivalence relation. $\Leftarrow$ (Right to Left).

We need to show now that R follows the symmetric and transitive properties since we know that it is already reflexive.

Check that it is symmetric.

Consider any $(a, b) \in R$. Since R is reflexive we have $(a, a) \in R$. Combining these two we get $(b, a) \in R$, by the property on the right of the iff. So it is symmetric.

Check that it is transitive.

Let $(a, b), (b, c) \in R$. Then $(b, a) \in R$, since we showed that R is symmetric. Combining this with $(b, c) \in R$ and using the property, we get $(a, c) \in R$. So it is transitive.

Since we have R to be reflexive and shown it is symmetric and transitive also, R satisfies the definition of equivalence relation. $\qquad\square$

Theorem 5.4.2 *An ordered pair for objects a, b, written $(a, b) = \{\{a\}, \{a, b\}\}$ (Kuratowski' definition). We have $(a, b) = (c, d)$ if and only if $a = c$ and $b = d$.*

Proof
$\Leftarrow$ (Right to Left)
Since $a = c$ and $b = d$ then forming the ordered pair $(a, b) = (c, d)$, by substitution.

$\Rightarrow$ (Left to Right)
We take $(a, b) = (c, d)$ and for a contradiction assume $a \neq c$ or $b \neq d$ or both. We will assume it is the first case for now. The proof should be the second case follows the same argument pattern as the first. Should it be the third case, then that is taken cared of by the first and the second situations.

By set equality definition let $X \in (a, b)$, thus $X \in \{\{a\}, \{a, b\}\}$.
Then there are two possibilities, either Case (I) $X = \{a\}$ or Case (II) $X = \{a, b\}$.

Case (I)
By definition of set equality, we have $X \in (c, d)$ also. For this situation, $X \neq \{c\}$ because we said that $a \neq c$. Thus $X = \{c, d\}$. We note that X is a set. So let $y \in X$, then $y = a$ since $X = \{a\}$. Also $y \in X$ implies $y \in \{c, d\}$, and since $a \neq c$, then $y = d$. Thus we have $a = y = d$. Therefore $(a, b) = (d, d) = (c, d)$. By set definition, this will imply that $c = d$ and that $a = d = c$, i.e., $a = c$. Contradiction.

Case (II)
We have $X \in (c, d)$. So either Case (a) $X = \{c\}$ or Case (b) $X = c, d$.

Case (a)
Let $y \in X$, then $y = c$. Consider then our Case II, i.e., $y \in X = \{a, b\}$. This means that $y = b$, since we know y cannot be equal to a. Hence, we have $c = y = b$. We have then $\{a, b\} = \{a, c\} = \{c\}$. By set definition $\{a, c\} \subseteq \{c\}$. Since $a \neq c$, we have an element on the left side of $\subseteq$ not a member of the set on the right. Contradiction.

Case (b)
Since by Case II, $X = \{a, b\}$ and $X = \{c, d\}$ then $\{a, b\} = X = \{c, d\}$. Since $a \neq c$, it must be that for $\{a, b\} = \{c, d\}$ we have $a = d$ and $b = c$. Then we have $(a, b) = (c, d) = (d, c)$. In order for $(c, d) = (d, c)$, by set definition, we have $c = d$. But $d = a$ which makes $c = a$. Contradiction.

Since for all possible cases we arrive at a contradiction, it must not be the case that $a \neq c$, thus $a = c$. Thus we have $(a, b) = (c, b) = (c, d)$, by substitution. By definition of ordered pair we have $\{c, b\} = \{c, d\}$ and by definition of set equality, this implies that $b = d$, as required. $\therefore$ in summary $a = c$, $b = d$.

Proof for the case $b \neq d$ is similar and will arrive at a contradiction. Proof, when both are the case (the third possibility), will also arrive at a contradiction since this first case is just a special case of the third case.   □

5.5   The Case Method

If you notice above, we spoke about cases, didn't we? In a way the previous discussion we actually did this method already, so in a way we "shot two birds with one stone".

Referring to the above, assume we have a proposition that is of this form—$P_1 \vee P_2 \vee P_3 \vee \ldots P_n \to Q$. To prove such a proposition we need to show that each P_i leads to Q. Each P_i is a case. Sometimes the cases are underneath a simple statement P in $P \to Q$. Thus, we deal with each case one by one and prove the result each time.

Theorem 5.5.1 $\max(x, y) + \min(x, y) = x + y$.

Proof There are two possibilities, either $x < y$ or $y < x$. If $x < y$ then $\max(x, y) = y$, $\min(x, y) = x$, then $\max(x, y) + \min(x, y) = y + x = x + y$. If $y < x$, $\max(x, y) = x$, $\min(x, y) = y$, then $\max(x, y) + \min(x, y) = x + y$. In all cases we get $x + y$.   □

In what follows we will use a known and previous theorem to illustrate that mathematics is built upon previously discovered propositions, i.e., theorems, to prove new ones. So a theorem steps on the shoulders of previous ones, just like what some people say about mathematicians—they stand on the shoulders of giants. It also illustrates that mathematics depends on the cooperative activity of its practitioners for its progress. The theorem we will refer to is the Euclidean Division Theorem (also known as The Division Algorithm [1]).

Theorem 5.5.2 (Euclidean Division Theorem) *In dividing integer a by a positive integer b, we can express $a = bq + r$, such that q, r are unique and that $0 \leq r < b$.*

Using the Euclidean Division Theorem we can prove the following statement which we will also use in the following example.

Theorem 5.5.3 *Every positive integer can be expressed in either 3 forms— $3k, 3k + 1, 3k + 2$ for some k and that the squares of these can be expressed in these forms—$3c, 3c + 1, 3c + 1$, respectively, for some c.*

Using the Euclidean Division Theorem in the above, it is not hard to see that a is any integer of concern and that $b = 3$ and $q = k$ whereas $r = 0, 1, 2$, the remainder when any integer is divided by 3. The property of their squares is obtained by algebraically manipulating the 3 forms as they are squared.

Theorem 5.5.4 *Assume that n is an integer then $n^3 - 2n$ is divisible by 3.*

Proof By Theorem 5.5.3, n can be expressed as either $3k, 3k + 1, 3k + 2$. If $n = 3k$, we have $n^3 + 2n = n(n^2 + 2) = 3k((3k)^2 + 2)$. This is divisible by 3 since clearly it is a factor of this product. If $n = 3k + 1$ we have $n^3 + 2n = n(n^2 + 2) = (3k + 1)((3k + 1)^2 + 2) = (3k + 1)(9k^2 + 6k + 1 + 2) = (3k + 1)(3)(3k^2 + 2K + 3)$. This again is divisible by 3. This last equation is divisible by 3 as it is again a factor. If $n = 3k + 2$ we have $n^3 + 2n = n(n^2 + 2) = (3k + 2)((3k + 2)^2 + 2) = (3k + 2)((3k + 2)^2 + 2) = (3k + 2)(9k^2 + 6k + 6) = (3k + 2)3(3k^2 + 2k + 2)$. Again the same as the previous result. Thus in all cases the polynomial is divisible by 3. The reader can look also at [1]. □

5.6 Mathematical Induction Method

We come now to one of the most confusing proof techniques that trips up students, the so-called proof by *mathematical induction*. This proof method is counter-intuitive but it is true. It is rather hard to wrap one's head around it yet it has been with us now for thousands of years. This technique pertains to proving a certain mathematical property holds for a mathematical object. For example, we might be asked to prove that every even integer >4 can be expressed as $p_1 + p_2$, where p_1, p_2 are prime numbers.

5.6.1 Weak Form

Here is our template on how to use this method. First show that the property P (like the property of being divisible, being prime or equal to a value, you name it, etc.) is true for the base case and the base case is the bottom value. If the base case is c then we prove that the property is true for it. So we work on showing $P(c)$ is true. Second, we assume it is now true for k in general, that is, $P(k)$ is true—we take this as fact. This second step is called the *induction hypothesis*—I.H. Lastly, we prove that $P(k + 1)$ is true as well using I.H. Thus, should we succeed in this final step, we have all the right to claim that we have proof that P is true for all n.

So in the above example, P is the property of being able to express an even integer >4 as the sum of two prime numbers. The base case is 6, the first even number greater than 4. The template says we should show that P is true for 6. Then we assume that it is true for k—any even number. After that, show it is true for the next even number, $k + 1$.

It is important that I.H. is used in the argument. Neglecting to use I.H. in the proof makes the proof invalid. This form of mathematical induction is called by some textbooks as the *weak form* of mathematical induction.

In the example below we shall apply this proof to a statement about *lucky numbers*. A number is a lucky number iff the function $L_m(n) = n^2 - n + m$ returns prime numbers for values of n, an integer, in $0 \leq n \leq m - 1$ [1].

Theorem 5.6.1 *The function $p(n) = n^2 + n + m$ returns prime numbers for all lucky numbers m, for $n = 0, 1, 2, \ldots (m - 2)$.*

Proof For $n = 0$, we get $p(0) = m$, but since m is lucky then by $L_m(0) = m$ and by definition m is prime too thus $p(0)$ is prime. I.H.: assume now that for $k \leq m - 2$, $p(k)$ returns a prime. We will show that $p(k + 1)$ returns a prime number too. From $p(k) = k^2 + k + m$ we get $p(k + 1) = (k + 1)^2 + (k + 1) + m = k^2 + 3k + 2 + m = k^2 + 4k + 4 - k - 2 + m(k + 2)^2 - (k + 2) + m = L_m(k + 2)$. Since m is a lucky number, we know that $L_m(k + 2)$ returns a prime number by definition thus $p(k + 1)$ returns a prime number too. $\square$

5.6.2   Strong Form

Sometimes in the course of the proof, we may need to appeal to the truth that the property is satisfied at a value r which is less than or equal to k. For example, the proof may need that the property is true for $(k - 1)$, $(k - 2)$ and so on. This is called the *strong form* of mathematical induction and this is when we assume $P(r)$ is true for the values of $r \leq k$. This I.H. assumes that the property P is true for values that are less than or equal to k.

We take the following from [1] once again.

Theorem 5.6.2 *If we have $a_0 = a_1 = a_2 = 1$ and $a_n = a_{n-2} + a_{n-3}$ for $n \geq 3$, then $a_n \leq (\frac{4}{3})^n$.*

Proof For the base case $n = 0$, we have $a_0 = 1 \leq (\frac{4}{3})^0 = 1$ which is true. For $n = 3$, we have $a_3 = a_1 + a_0 = 1 + 1 = 2 \leq (\frac{4}{3})^3$. For I.H. we assume $a_k \leq (\frac{4}{3})^k$ and for values less than equal to k. We have $a_{k+1} = a_{k-1} + a_{k-2}$. By I.H. $a_{k-1} \leq (\frac{4}{3})^{k-1}$ and $a_{k-2} \leq (\frac{4}{3})^{k-2}$. Adding these two we get $a_{k+1} = a_{k-1} + a_{k-2} \leq (\frac{4}{3})^{k-1} + (\frac{4}{3})^{k-2} = (\frac{4}{3})^k((\frac{4}{3})^{-1} + (\frac{4}{3})^{-2}) \leq (\frac{4}{3})^k(\frac{4}{3}) = (\frac{4}{3})^{k+1}$. $\square$

In the above proof, notice that we appealed to the truth of the property at $k - 1$ and $k - 2$.

5.6.3 Why It Works

There are a few comments that we can make why mathematical induction (MI) works:

Firstly, MI is about the property of numbers (in general). Numbers obey this induction property. As a classic example consider a number x such that $x > c$, for some number like 8. So if we have $x > 8$, can we say that the next number after x will also be greater than 8? We can guess that this should be true. So let us prove it. Consider $x > 8$, let us add 1 to both sides still maintaining the inequality. $x > 8 \Rightarrow x + 1 > 8 + 1 = 9 > 8 \therefore x + 1 > 8$.

Secondly, we are allowed to assume I.H. for after all we can set our $n = c$, i.e., to our base case and check if the property $P(c + 1)$ holds—thus by the same token we are allowed to move from the truth of $P(c)$ to the truth that $P(c + 1)$. So we can assume I.H. because of this "domino effect". The crucial bit is to show now that due to our utilization of I.H. for an arbitrary $n = k$, we get the truth of $P(k + 1)$.

Lastly and strongly, we can take MI to be an axiom! Meaning, a proposition which is self-evidently true! Indeed Wikipedia has it like this—$\forall P[P(0) \wedge \forall k \in \mathbb{N}[P(k) \Rightarrow P(k + 1)]] \Rightarrow \forall n \in \mathbb{N}[P(n)]$.

Take a good look at this and consider the statement before the last $\Rightarrow$, the one enclosed by [. . .]. If you look, we have this form $A \Rightarrow B$. Remember *modus ponens*? It says if we have $A \Rightarrow B$ and we have A, deduce B. Well when we are doing MI what we are actually doing is that we are establishing the truth of A and when we succeed—voila!, we use this with the axiom and so conclude the property P holds for all n.

We can also apply MI to statements that, on the surface, do not appear to be about numbers. Here are a few examples.

Theorem 5.6.3 *The number of subsets of a set of n elements is equal to 2^n. Notationally, let A be a set with n elements then $|\wp(A)| = 2^n$. (Note the symbol $|$ that surrounds $\wp(A)$ stands for the number of elements in a set.)*

Proof For the base case we let $A_1 = \{a_1\}$, i.e., one element, then $\wp(A_1) = \{\{a_1\}, \varnothing\} \Rightarrow |\wp(A_1)| = 2 = 2^1$. So it is true for the base case.

For I.H. we assume that if $A_k = \{a_1, a_2, a_3, \ldots, a_k\}$, then $|\wp(A_k)| = 2^k$. Our target is to show that for $A_{k+1} = A_k \cup \{a_{k+1}\}\}$, we have $|\wp(A_{k+1})| = 2^{k+1}$.

Consider the subsets of A_{k+1}, we can divide this to the subsets that do not contain a_{k+1} and the subsets that will contain a_{k+1} The subsets of A_{k+1} that does not contain a_{k+1} are the subsets of A_k, and by I.H. we know this to have 2^k subsets. Consider now the subsets of A_{k+1} that will contain a_{k+1}, then this is formed by taking a subset of A_k and chucking in a_{k+1} into it. Then by I.H. again, the number of subsets that will contain a_{k+1} is 2^k. Thus forming

$$\wp(A_{k+1}) = \wp(A_k) \bigsqcup [\bigcup_{s_i \subset A_k} s_i \cup \{a_{k+1}\}]$$

$$|\wp(A_{k+1})| = |\wp(A_k)| + |[\bigcup_{s_i \subset A_k} s_i \cup \{a_{k+1}\}]|$$

$$|\wp(A_{k+1})| = 2^k + 2^k = 2 \cdot 2^k = 2^{k+1}$$

which is required.   □

Do not get scared of the symbols found in the above proof. I just used them to be more economical with my words. The bit symbolized by $\sqcup$ stands for union also and

$$\bigcup_{s_i \subset A_k} s_i \cup \{a_{k+1}\}$$

stands for collecting all subsets of A_k (one of which is an s_i) where a_{k+1} is inserted in it.

MI can also be applied to what appears to be a statement in geometry.

Theorem 5.6.4 *Suppose we lay down n lines on a plane wherein none of them are parallel and no 3 of them have a common point, then the number of regions cut by these lines is $\dfrac{n^2 + n + 2}{2}$.*

Proof The base case when we only have one line is shown in Fig. 5.1. As we can see we have 2 regions. This comes from $(1^2 + 1 + 2)/2 = 4/2 = 2$. To help our imagination we also have Fig. 5.2 which clearly shows we have $\dfrac{2^2 + 2 + 2}{2} = \dfrac{4 + 4}{2} = 4$.

For I.H. assume now we have n lines and the number of regions is $|R(n)| = \dfrac{n^2 + n + 2}{2}$. Our target is to show that $|R(n+1)| = \dfrac{(n+1)^2 + (n+1) + 2}{2}$. So consider now adding one line to $R(n)$, so the new number of regions would be the existing number of regions plus the new regions contributed by the cutting of the new line on existing lines. So we have $|R(n+1)| = |R(n)| + m$, where m is number of regions contributed by the new line. We can observe that this new line will cut existing lines and will add 1 region for every line it cuts plus 1. Since it will cut n lines the number of regions contributes will be $m = n + 1$. Hence we have

$$|R(n+1)| = |R(n)| + m$$

$$= |R(n)| + (n+1) = \frac{n^2 + n + 2}{2} + (n+1)$$

$$= \frac{n^2 + n + 2 + 2n + 2}{2} = \frac{n^2 + 2n + 1 + (n+1) + 2}{2}$$

$$= \frac{(n+1)^2 + (n+1) + 2}{2}$$

This completes our induction step and shows what is required. □

5.7 Reflections

Reflection 5.7.1 *Check that p in Theorem 5.6.1 does indeed return a prime number when n is equal to its upper bound.*

Reflection 5.7.2 *Consider a number composed of all 1s. There is a theorem that says that if the length of this number is equal to 3^n, then this number is divisible by 3^n. For example 111 is divisible by 3^1, i.e., $111 = 3 \times 37$. Consider also $111, 111, 111 = 3^2 \times 12345679$, and so on. See if you could prove this theorem using mathematical induction.*

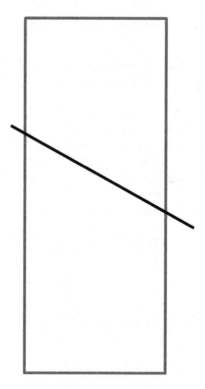

Fig. 5.1 When $n = 1$

5.8   Proof by Construction

This proof method is used to demonstrate usually the existence of a mathematical object which satisfies a certain property. You will usually find this type of proof in Geometry and also in Calculus, or what is called, Mathematical Analysis like Real or Complex Analysis. The proof outlines or spells out the process of finding the object. Occasionally, we can get away from proving existence using proof by contradiction. Sometimes this does not work in a smooth way, so to use this proof you have to provide mechanical actions that lead to what is required.

We can give an analogy of this in real like. Say someone asserts "Unicorns exists". To prove this, the one asserting simply brings in front of you the unicorn!

Theorem 5.8.1 *In plane geometry, given an arbitrary two points A and B on a plane, there is always a line $\overline{AB}$ that passes through these points.*

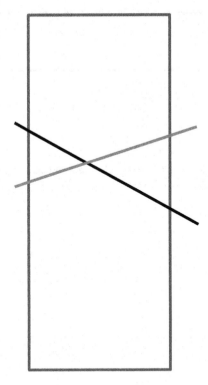

Fig. 5.2 When $n = 2$

Fig. 5.3 Proof by construction of a line from A to B

Proof Put two points anywhere on a plane and using a ruler make it pass through point A to point B. See Fig. 5.3. □

Comment

Notice how there was a process on how to see the existence of what is being asserted to exist.

Reflection 5.8.1 *Prove this: Let x be any integer, then there exists an integer y such that x < y.*

Reference

1. J.L. Mott, A. Kandel, T.P. Baker, *Discrete Mathematics for Computer Scientists and Mathematicians* (Prentic Hall, 1986)

Part II
An Application

In this part, we will have the opportunity to apply what we have learned in Part I. Here, we will use logic and mathematics to study logic itself. We will prove some theorems and thereby show more examples of how the lessons learned in Part I work in action. However we won't use logic to study logic for its own sake. We have a parallel goal, and this is to answer a single question: given a set of sentences, can we add an extra sentence, and keep the whole new set consistent? The answer to this question has many interesting real-life applications.

Chapter 6
Formal System for PL

> *"In a way, math isn't the art of answering mathematical questions, it is the art of asking the right questions, the questions that give you insight, the ones that lead you in interesting directions, the ones that connect with lots of other interesting questions-the ones with beautiful answers."*
>
> —George Chaitin

Abstract In this chapter we will actually do some mathematics but we will apply the mathematics into logic itself. In other words we will study logic from a mathematical point of view. We will use logic and mathematics to study logic itself. Scholars sometime call this exercise *metamathematics*. We will first apply this type of study into Propositional Logic (PL).

6.1 PL as a Formal System

In Part I, we covered two logical reasoning approaches, namely, Propositional Logic and First-Order Logic. The latter has a more micro-level analysis of statements while the former looks at statements which convey a complete thought, so more macro-level. As a formal science, logic investigates how valid statements can be derived from other statements. A *Formal System* is a way of looking at a logical system from a symbolic point of view in a systematic and abstract manner. It studies logic through the forms found in its formulas. This is where we get the term "formal". The tasks in a formal system are to develop precise mathematical theory that explains which inferences are valid and why.

If we notice, we have been doing some abstractions in Part I. Abstraction is the technique of extracting the underlying structures or properties of a mathematical concept by hiding unnecessary details of the subject under study. The purpose of which is to focus on the important details, get into the heart of the matter, and to understand well how the subject under investigation works. A formal system may be formulated for its intrinsic properties or it may be designed as a description of external phenomena. For example, in physics, we have classical mechanics. This can

© Springer Nature Switzerland AG 2021
L. P. Cruz, *Theoremus*,
https://doi.org/10.1007/978-3-030-68375-7_6

be thought of as a formal system as well. We can add more examples, but in the mathematical discipline these systems are also referred to as *Theories*.

Elements of a Formal System

A formal system needs several elements:

1. A set of propositional symbols from which we can create more complex formulas.
2. A rule to say when is a formula is properly formed or written .
3. Some set of axioms which are also in symbolic form.
4. A set of rules of inference or deduction rules for producing more new formulas.
5. The exercise of the above produces new formulas which are called theorems. .

Some authors do not include Point 5 because that is the result of the whole process of abstraction. However, we include it here because when we refer to a mathematical system we refer to its results, thus their theorems too. Remember again that these systems are called theories, for example, Number Theory, Set Theory, Probability Theory, etc.

6.2 PL Syntax

In this section we will proceed with the systematization of PL, bearing in mind the lessons learned in Part I, Chaps. 3 and 4.

First, let us the building blocks of propositional formulas, called *propositional variables*. Propositional variables are the *primitive or atomic* propositions in PL. To say that p is a primitive/atomic proposition means it is the simplest proposition. It is not a compound statement made of several clauses.

Let us now cover Point 1 in Sect. 6.1 which is about symbols and formulas. Mathematicians use a symbol for the phrase "defined as". They sometimes use $\equiv$ or $\stackrel{\circ}{=}$ to say that what is on the left-hand side of the symbol is defined as the one on the right-hand side. For example, let $p \stackrel{\circ}{=}$ "x is an even number". Observe that as a statement, it is simple, i.e., it is not a compound statement and cannot be broken down to any constituent parts; it does not have any connectives.

The next element we present is the syntax of PL covering Point 2. Syntax refers to the grammar of a language. It is the set of rules that say how sentences are formed. In logic this is done using a syntax diagram, which we will show below. This diagram visually teaches how correct formulas or statements are constructed.

Figures 6.1 and 6.2 show the syntax diagram for PL. It will show how a PL formula, say F, may be formed. We follow the arrows and they show us how to make more complex statements from the atomic ones. We define F iteratively. In forming the formula, we simply follow the rules presented by the arrows. Some authors call them *well-formed formulas*.

So let us explain how to read it. Let us go first to the definition of *atomic* then we will deal the larger context of F.

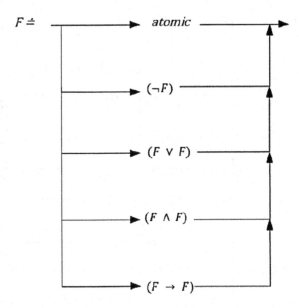

Fig. 6.1 Syntax diagram of PL formulas

Here an *atomic* proposition is a proposition that cannot be broken into several parts anymore. For example, we may say "Tommy is a boy", "The cat sat on a mat", "The quick brown fox jumped over the lazy dog", etc. We may represent this with a proposition small letter such as p or q or any small letter we can supply. When we run out of propositional letters due to the many assertions we are making, we can use subscripts such as p_1, p_2, and p_3. The symbols $\top$, $\bot$ stand for any generic TRUE or FALSE statement. The word *constants* come to mind. A constant is not a propositional variable. Recall that in mathematics we have constants, like π, e, i, etc. We have the same concepts in logic as well. This idea of constants will become clearer later in the sections that will follow.

Examples

We can verify that the examples below are well-formed by applying the rules presented by the syntax diagram. To verify, we ask if we can fit the said formula into the syntax diagram rule?

1. $(p \vee q)$
2. $((p \wedge (\neg q)) \to t)$
3. $((\top \wedge p) \vee (r \to q))$
4. $\neg(\neg q)$ is not well formed by our rules. It should be $(\neg(\neg q))$.

Convention

In the fourth example above we could have made that statement true to our syntax diagram, by the use of parenthesis. We notice that our syntax diagram rule is quite

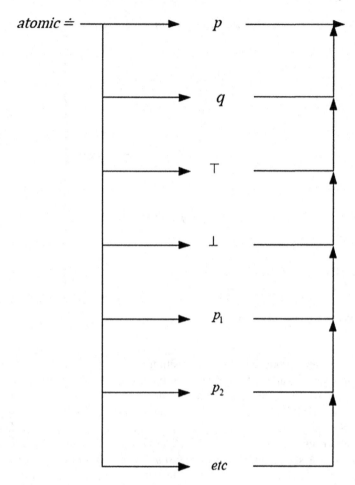

Fig. 6.2 PL atomic formulas

pedantic and explicit. The parenthesis makes sure we know how they are being grouped together with what is known as *connectives*, namely, $\neg$, $\vee$, $\wedge$, $\rightarrow$. The syntax diagram specifies how formulas bind together. When there is no risk of confusion and for economy, we can drop the parenthesis when the meaning is clear. Usually, this has something to do with the outer parenthesis. For instance we can simply say $\neg p$ rather than $(\neg p)$ or $(p \wedge (\neg q)) \rightarrow t$ rather than $((p \wedge (\neg q)) \rightarrow t)$.

The binding priority and convention is that first $\neg$ binds to its formula to the right, then both $\vee$ and $\wedge$ bind to the formula to their left and right, finally $\rightarrow$ binds to the formula before and after it. Thus for $(p \wedge (\neg q)) \rightarrow t$, we can further economize by stating its equivalent of $(p \wedge \neg q) \rightarrow t$, and finally even to this $p \wedge \neg q \rightarrow t$. The parenthesis helps us to be more explicit and communicate to the reader which formula takes which connective it belongs.

We will also have the following equivalence:

$$\neg F \equiv F \rightarrow \bot$$

Examples

Let's do some analysis to find out if something is a well-formed formula.

1. Looking back at $\neg\neg p$. Is this a correct formula? Using the convention we get $\neg(\neg p)$ because of the $\neg$ binding rule.
2. The symbol $(p \times q)$ is not a valid formula since $\times$ is not part of our symbol.
3. The symbol $((p))$ cannot fit in any formation rule specified to us. If we want to say something like this, it is enough to simply assert p.
4. Similarly rst is not a legal formula; we cannot fit it into any of our formation rules.
5. Lastly $p \wedge q \vee r$ is legal. Using the convention rules, we get $(p \wedge q) \vee r$.
   Consider the left most part of this symbol, $(p \wedge q)$. We know p, q, r are atomic and thus $(p \wedge q)$ follows the formation rule and hence, this is a formula too. We will use the symbol $\rightsquigarrow$ to notate "becomes". We will use it to say when a formula is transformed to another formula based on the rules. Thus we get $(p \wedge q) \vee r \rightsquigarrow ((p \wedge q) \vee r)$ by convention. We see then that following the convention and the form of the rule we conclude that the original $p \wedge q \vee r$ is a formula.

When see formulas of $A \vee B$ form, then this is called a *disjunction* of A and B. In the same way, when we see formulas of the form $A \wedge B$, we say that this is the *conjunction* of A and B.

Notation

Later on in the following sections we will find $(\ldots((A_1 \vee A_2) \vee A_3) \vee \cdots \vee A_n)$. This is shortened by

$$\left(\bigvee_{i=1}^{n} A_i \right)$$

and A_i is of course a formula.

Likewise when we find $(\ldots((A_1 \wedge A_2) \wedge A_3) \wedge \cdots \wedge A_n)$. This is shortened by

$$\left(\bigwedge_{i=1}^{n} A_i \right)$$

Below is a challenging question that deserves deeper examination. Use your reasoning powers to answer the question.

Reflection 6.2.1 *Is* $\neg\neg p \vee q$ *a formula? Using the convention and rules put back the parenthesis on this formula. Is this different from* $\neg\neg(p \vee q)$? *Is this then a formula? Form an argument as to why or why not it is a formula.*

6.3   Induction on PL Formula

In the previous parts of this book we discussed mathematical induction. In this section we will introduce something similar but this pertains to logical formulas. This is known as *proof by induction on the formula*. Other authors call this method *structural induction* because it looks at the structure of the formula. Technically we should have discussed this in Part I but we would not be ready for this because we, at that time, had not spoken about forms of formulas.

In simple terms it says that in any formal system, a formula in it will have a certain property Ψ if we can show that each formula, produced according to the formation rule, has property Ψ. A property of a formula is any characteristic description we can ascribe to it. For example, we can say, any formula has a balaned left and right parenthesis. Such a statement is a property etc. Hence, we can designate the said property as our Ψ.

Theorem 6.3.1 (Structural Induction) *Let* Ψ *be a property. Let* F *be any formula in PL. To show that every* F *obeys property* Ψ, *i.e.,* $\Psi(F)$ *it is enough to show the following hold true:*

1. $\Psi(p)$, *for any* p *in PL.*
2. *If* $\Psi(F)$, $\Psi(\neg F)$.
3. *If* $\circ$ *stands for* $\vee, \wedge, \rightarrow$, *then* $\Psi(F \circ G)$, *if* F, G *such that* $\Psi(F)$ *and* $\Psi(G)$.

Proof Proving by induction.
Base Step: p is any atomic formula then we should show that $\Psi(p)$. Note that p has no or zero connectives. That takes care of the first point above.
Induction Step: Let us assume that any formulas F, G have n connectives and let us assume that $\Psi(F)$ and $\Psi(G)$. We need to show that Ψ should hold for formulas that have $(n + 1)$ connectives.

Consider $(\neg F)$,this has $(n + 1)$ connectives since F has n, for here we just added the connective $\neg$. So it is enough to show that $\Psi(\neg F)$. That takes care of the second point above.

Consider now the formula $(F \circ G)$, this formula is like the one above in that we are just adding one other connective to the n connectives we have together from F and G. The connective added is the one that stands for $\circ$. So thus it is enough to show that $\Psi(F \circ G)$. $\qquad\square$

Reflection 6.3.1 *If we re-read the theorem once again, we note that it is a statement on how to prove. It is a kind of a* meta-theorem. *"Meta" is the Greek word for being "beyond" or "above" one's self, some kind of self-awareness. Consider that it is a theorem on how to prove, essentially, a theorem in PL. Is this reasonable? Explain why.*

Comment

The theorem in PL we are discussing here is about properties of formulas and of course, when we say such and such formula F follows property Ψ, that effectively is a theorem, an assertion begs to be proved. We are of course making an assertion.

Example

Let consider a very common way of illustrating the application of structural induction; that is to show that every formula F has a balanced left and right parenthesis—Ψ, i.e., (), or having an even number of matching parenthesis, based on our formation rules, i.e., Fig. 6.1.

Theorem 6.3.2 (Metatheorem Example) *Every valid F has a balanced left and right parenthesis.*

Proof First, an F can be a p. It has no paired parenthesis or zero parenthesis thus even. One may argue it has no parenthesis thus also balanced. So $\Psi(p)$. This is our base case.

Now we need to consider for all other connectives. For our IH, we assume that F and G have balanced parenthesis in them.

If F and G have balanced pairs of parenthesis in them internally, we see that $(\neg F)$ and $(F \circ G)$ do have again a balanced pairs because it required us to put a left (and a right) in the formation rule. Hence, it is the case that $\Psi((\neg F))$ and $\Psi((F \circ G))$ where $\circ$ are our disjunction, conjunction and implication connectives. □

Reflection 6.3.2 *See if you could apply the above to prove the property Ψ: Every F in PL is composed of atomic formulas. To some this is obvious but for practice we can be pedantic and mechanically prove this anyway.*

6.4 PL Semantics

The semantics of a formal system are the rules of how we interpret or give meaning to its syntax or symbolism. If we are to model statements so we can reason with them, it is not enough to have syntax, we must have semantics as well. In semantics we are concerned with the TRUTHFULNESS or FALSITY of a sentence which is expressed in the symbolic syntax given in the said formal system.

We have seen this already in Sect. 4.1.2 when we spoke about truth tables. We shall come back to it here but this time we will give a more algebraic way of dealing with the meaning of a PL formula. This time we will use a function to define the TRUTHFULNESS or FALSITY of a PL statement. Just like as before we can determine this through an iterative definition of the syntactical component parts; thus the overall identification of the sentence is true or false that depends on the state of the atomic component parts. We will call such an identification an *interpretation*.

Definition 6.4.1 (PL Interpretation) *The Interpretation for PL is a function* $[[\bullet]]^I : PL \rightarrow \top, \bot$ *under the following rules, where* $\bullet$ *is a PL syntactic statement and that* $\min(\top, \bot) = \min(\bot, \top) = \bot$ *and* $\max(\top, \bot) = \max(\bot, \top) = \top$:

1. $[[\bot]]^I = \bot$.
2. $[[\top]]^I = \neg[[\bot]]^I = \top$.
3. *With P and Q are formulas,* $[[P \wedge Q]]^I = \min([[P]]^I, [[Q]]^I)$.
4. $[[P \vee Q]]^I = \max([[P]]^I, [[Q]]^I)$.
5. $[[P \rightarrow Q]]^I = \bot$ *iff* $[[P]]^I = \top$ *and* $[[Q]]^I = \bot$.
6. $[[P \leftrightarrow Q]]^I = \top$ *iff* $[[P]]^I = [[Q]]^I$ *or* $\bot$ *otherwise.*
7. $[[\neg P]]^I = \neg[[P]]^I$.

Convention

It is customary to simply use $[[\bullet]]$ without any decorations. In our case, we will drop the I superscript whenever the meaning is clear. See for example [1].

Remark

P, Q may stand for atomic or compound formulas. The notation $[[P]]$ should either map to $\top$ or $\bot$ depending if P matches its reality or not. For example, let $P \equiv$ "John is a father", then $[[P]]$ is $\top$ if indeed the John we are talking about has children, otherwise the value of $[[P]]$ is $\bot$.

In the above we need to think more deeply about how we defined the truthfulness of $P \rightarrow Q$. Note that we have defined when it is FALSE, so the negation of that is when it is TRUE.

Example

Let us show that $p \lor \neg p$ is always $\top$ (TRUE) for any interpretation. Intuition says that this is always the case, but let us convince ourselves of this truth
Let us calculate
$[[p \lor \neg p]] = \max([[p]], [[\neg p]])$
$\Leftrightarrow = \max([[p]], \neg[[p]])$
$\Leftrightarrow = \max([[p]], \neg[[p]])$
$\Leftrightarrow = \max(\top, \neg\top) = \max(\top, \bot) = \top \Leftrightarrow [[p]] = \top$
or
$\Leftrightarrow = \max(\bot, \neg\bot) = \max(\bot, \top) = \top \Leftrightarrow [[p]] = \bot.$

So either way, whatever p might be it will always be the case that $p \lor \neg p$ is always TRUE.

Example

In the same vain, let us show that $p \land \neg p$ is always $\bot$ (FALSE) for any interpretation. Likewise, intuition says that this is always the case, as usual, let us convince ourselves of this truth
Let us calculate
$[[p \land \neg p]] = \min([[p]], [[\neg p]])$
$\Leftrightarrow = \min([[p]], \neg[[p]])$
$\Leftrightarrow = \min([[p]], \neg[[p]])$
$\Leftrightarrow = \min(\top, \neg\top) = \min(\top, \bot) = \bot \Leftrightarrow [[p]] = \top$
or
$\Leftrightarrow = \min(\bot, \neg\bot) = \min(\bot, \top) = \bot \Leftrightarrow [[p]] = \bot.$

So either way, whatever p might be it will always be the case that $p \land \neg p$ is always FALSE.

Reflection 6.4.1 *Can you show that $(p \land q) \rightarrow p$ is always $\top$? Consider that since we have 2 propositional atomic symbols, then there are 2^2 possible combinations of $\top$ and $\bot$. Do you think using truth tables we have encountered in Part I might be the easiest way to prove this?*

Below we have a theorem that is quite useful when it comes to semantics and most authors like [1, 2] consider it important not to be left out. However, we take inspiration from [2].

Theorem 6.4.1 *Let F be a formula, and let I and J be interpretations on F. Let $[[p]]^I = [[p]]^J$ for all atomic proposition $p \in F$. Then $[[F]]^I = [[F]]^J$.*

Proof The first condition on atomic propositions already holds in that it is given we have $[[p]]^I = [[p]]^J$.

We look at the form $F = F_1 \vee F_2$. By inductive hypothesis (IH) J and I agree on their interpretation on F_1 and F_2. We need to show that $[[F_1 \vee F_2]]^I = [[F_1 \vee F_2]]^J$. Let us form $\max([[F_1]]^I, [[F_2]]^I) = \max([[F_1]]^J, [[F_2]]^J)$ by IH $\Rightarrow [[F_1 \vee F_2]]^I = [[F_1 \vee F_2]]^J$ as required. The reader can follow through on other forms and can convince him/herself that this is true for all forms. $\square$

Recall that in the above example we have a formula $p \vee \neg p$ which is always $\top$. This means forming the negated disjunction of a proposition with itself is always results in a TRUE statement. Logicians have a word for this, they call this case a *tautology*. With these formulas, no matter what truth/false assignment you make to its components they wind up always TRUE. So they are independent of the interpretation.

Definition 6.4.2 *Let F be a formula in PL. Then F is called a* tautology *iff for any interpretation J, $[[F]]^J = \top$.*

Example

$p \rightarrow p$ is a tautology, and you can convince yourself of this fact.

Likewise $(p \rightarrow q) \rightarrow (p \rightarrow q)$ is a tautology. You can convince yourself again of this fact. You really do not have to do the truth table here. You can detect that it is a tautology because it follows the form of the first one which we know is a tautology already. Recall the idea of *schemas* in Sect. 4.1.3.

Definition 6.4.3 *An algorithm $\mathcal{A}$ is a finite set of actions or procedures $a_1, a_2, \ldots, a_n$ that are unambiguously defined such that an agent (human/machine) can mechanically perform them.*

Notice that we used the word "mechanical" in the above definition. Computer scientists call it an effective procedure because an algorithm obviously has to finish its work and produce the correct answer.

Example

1. The recipes we find in a cook book are algorithms.
2. We can make an algorithm for making coffee.
3. When we ask our GPS navigation device to give us directions from A to B it actually gives us an algorithm for traveling from A to B.

Reflection 6.4.2 *Could you name a couple of algorithms you find in our world?*

Theorem 6.4.2 *For every F formula in PL, there is an algorithm to determine if it is a tautology or not.*

Proof The algorithm is the construction of the truth table for F. □

6.5 Satisfiability, Validity and Consequences

Given an arbitrary statement, would it not be a great idea to find out if that statement is true, or false? That would be a great knowledge to have. The above concepts will give us ideal properties that will help in pointing the way to such a quest. We already talked about a part of this when we spoke about tautologies.

Definition 6.5.1 *Let F be a formula in PL. Then we say F is* satisfiable *if there is at least one $\mathcal{J}$ so that $[[F]]^{\mathcal{J}} = \top$. We also say that F holds under interpretation $\mathcal{J}$. If we cannot find such an interpretation $\mathcal{J}$ that can make F TRUE then we say that F is* unsatisfiable. *We may also say that when an interpretation can make F FALSE then the formula is* falsifiable.

In the following theorem, we will use what we learned in Sect. 5.3.

Theorem 6.5.1 *T is a tautology iff $\neg T$ is unsatisfiable.*

Proof $\Rightarrow$
Assume that T is a tautology and $\neg T$ is satisfiable. T is a tautology means that every interpretation like $\mathcal{I}$ implies that $[[T]]^{\mathcal{I}} = \top$. If $\neg T$ is satisfiable, then by definition, there is at least one $\mathcal{J}$ such that $[[\neg T]]^{\mathcal{J}} = \top$. Which means by definition $[[\neg T]]^{\mathcal{J}} = \neg[[T]]^{\mathcal{J}} = \top$. But since T is a tautology then $[[T]]^{\mathcal{J}} = \top$ which implies that $\neg[[T]]^{\mathcal{J}} = \neg\top = \bot$, so this implies too that $[[\neg T]]^{\mathcal{J}} = \bot$, a contradiction.
$\Leftarrow$
We can show this by an analogous argument. Reader, please try this yourself. □

Reflection 6.5.1 *Try proving the $\Leftarrow$ of Theorem 6.5.1.*

Definition 6.5.2 *We write tautologies like T as* $\models T$, *if it is not then we write it as* $\not\models T$. *We also say T is* valid *if it is a tautology.*

Definition 6.5.3 *Let* $\Gamma = \{F | F \in PL\}$, *we say that* Γ *is* satisfiable *iff there is at least one interpretation* $\mathcal{I}$ *such that for all* $F \in \Gamma$ *we have* $[[F]]^{\mathcal{I}} = \top$. *We also say* $\mathcal{I}$ satisfies Γ *and that* $\mathcal{I}$ *is a* model *for* Γ. *If we cannot find an interpretation that satisfies* Γ *then it is* unsatisfiable.

Definition 6.5.4 *Let* Γ *be a set of formulas and let F be any formula. If every interpretation* $\mathcal{J}$ *satisfies* Γ *also satisfies F, then we say that F is a* consequent *of* Γ. *We symbolize this by writing it as* $\Gamma \models F$. *We also say that* $\Gamma \models F$ *is* valid.

To find out if $\Gamma \models F$, we need to show that for every $\mathcal{J}$ that satisfies each $A \in \Gamma$ also satisfies F. So, therefore, we simply test the situation against the definition above. Further, note that the set all together must be satisfiable.

Recall Definition 6.5.2. We note that $\models F$ is a special case of Definition 6.5.4, $\Gamma = \varnothing$, the empty set. Also notice that we have mentioned the concepts such as validity and satisfiability twice, one without any reference to a set of formulas Γ and the other, with indeed reference to Γ.

Why do we do this? Well we will learn that mathematicians are interested in Theories. For example, take the field of Number Theory. They want to discover new theorems that are true based on existing formulas $F_1, F_2, F_3, \ldots$, etc. These formulas constitute analogously the Γ we are talking about here.

We may view the relationship of these concepts using the diagram. See Fig. 6.3

In the diagram, we need to see that the circles represent logical formulas or statements. The green circle is statements that are tautologies. The lower part of it is an example of such a formula that is always TRUE. On the right, the red circle contains the formulas that are always FALSE no matter what kind of interpretation we have. The middle one, the one in brown are statements that may swing both ways depending on whether or not we find an interpretation that helps our purpose. For example we can have $p = $ "Felix is a cat". If there is such a cat in the world whose name is Felix, then this statement is $\top$. However, if it so happens we have a Felix who happens to be a boy, then this interpretation makes $p = \bot$, so this is false.

Let us do some proofs.

How to view the concepts

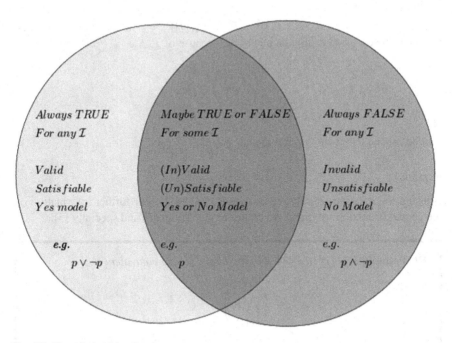

Fig. 6.3 The relationship of concepts

Theorem 6.5.2 *Let* Γ *be a set of formulas* $\{P_1, P_2, P_3, \ldots, P_n\}$*. Then* $\Gamma \models Q$ *iff*

$$\models \left(\bigwedge_{i=1}^{n} P_i \right) \to Q$$

i.e., they are equivalent.

Proof $\Rightarrow$

$\Gamma \models Q$ iff for all $\mathcal{I}$ means for all $P_i \in \Gamma$, we have $[[P_i]]^{\mathcal{I}} = \top$ implies too that $[[Q]]^{\mathcal{I}} = \top$ by definition of $\models$. Let us form $[[P_1 \wedge P_2 \wedge P_3 \wedge \cdots \wedge P_n]]^{\mathcal{I}} = \max(P_1, P_2, P_3, \ldots, P_n) = \max(\top, \top, \top \ldots, \top) = \top$. Let us form now

$$\left[\left[\left(\bigwedge_{i=1}^{n} P_i \right) \to Q \right]\right]^{\mathcal{I}}$$

This is TRUE as per Table 4.2 iff

1. the left of $\to$ is TRUE and the right of it is TRUE or

2. the left is FALSE and the right is TRUE or
3. both are FALSE

Since, $[[P_1 \wedge P_2 \wedge P_3 \wedge \cdots \wedge P_n]]^I = \top$ and $[[Q]]^I = \top$ it can only be the first option as the case and cannot be otherwise because of I. $\therefore$

$$\models \left(\bigwedge_{i=1}^{n} P_i \right) \to Q$$

$\Leftarrow$

Can be proved using similar argument. $\square$

Remark

Abusing the notation slightly we can write $\models \Gamma \to Q$. For further information on the above discussion we direct the reader to the references and specially [3].

> **Reflection 6.5.2** *Try to prove that the above is also equivalent to*
>
> $$\left(\bigwedge_{i=1}^{n} P_i \right) \wedge \neg Q$$
>
> *being unsatisfiable.*

6.6   PL Proof System

A logic will not be complete without the rules for deduction. We have encountered this already in Sect. 4.1.4. There we used Fitch's Style Natural Deduction. We introduced this style because it mimics the way mathematicians argue when they prove. This is the reason why logicians call it *natural deduction*—it complies with our intuition and naturally follows the path of argumentation. Granted, mathematicians do not always follow the linear descent we see in Fitch Style, they do not draw lines like we find there, most of the time as we have seen here too, they use prose in argumentation.

For the sake of keeping us all current we will introduce the reader to the so-called Gentzen Style Natural Deduction found below.

Fig. 6.4 Gerhard Gentzen

6.6.1 Gentzen Style ND

The Gentzen style of natural deduction came from Gerhard Gentzen a German mathematician who lived in the early part of the twentieth century, shown in Fig. 6.4.[1]

Gentzen proposed actually two forms of his proof style system. The one we mentioned is natural deduction presentation and the other is called *sequent calculus*. We won't deal with the latter and we leave it to the reader to read up on this in his/her own free time.

Why do we have to introduce Gentzen Style Natural Deduction? Well, it has the convenience of easy subtraction and addition of the universal and existential quantifiers which is found in Predicate Calculus or First-Order Logic, something talked about in Sect. 4.2. It reduces the derivation of PL to syntax manipulation. Furthermore, it allows suppositional assumptions in the proof the same way that

---

[1]Credits: Eckart Menzler-Trott, CC BY-SA 2.0 DE https://creativecommons.org/licenses/by-sa/2.0/de/deed.en, via Wikimedia Commons.

Fitch does. These are propositions we temporarily assume for the sake of argument and later on we discard. So looking at it now, helps us appreciate it later. For example, let us say we have the case that $P \vee Q$. Assume it is P, what can we later deduce from this then, and so on, etc.?

Gentzen is analogous to Fitch except that in Gentzen style we do our derivations, not straight down vertical but sometimes horizontal which we will see an example of this below.

Let us recall again the concepts we discussed in the previous chapters on inference rules. We will now give a formal definition of a proof in the Formal System PL by considering the derivation shown below:

$$
\begin{array}{c|c}
1 & E_1 \\
2 & E_2 \\
3 & E_3 \\
4 & \vdots \\
5 & E_n \\
\hline
6 & F
\end{array}
$$

Definition 6.6.1 *Then based on the above, each E_j, $j \leqslant n$ is either a premise, or the result of applying an axiom in PL or the result of an inference rule on some previous E_i, where $i \leqslant j - 1$. If such is the case then we say that $E_1, E_2, E_3, \ldots, E_n$ is a proof of F. When we wrap all the E_i into a set, i.e., $\Gamma = \{E_1, E_2, E_3, \ldots, E_n\}$ we say Γ proves F and we write $\Gamma \vdash F$. Further, we say that F is a theorem of Γ.*

6.6.2 Gentzen Inference Rules

Each author may present either more or less inference rules shown here. They may provide a few axioms or none at all [1, 2]. At any rate the approach is the same.

Below we show our version of these rules.

Definition 6.6.2 *We list here the rules of the Gentzen natural deduction we will use in this work.*

$$\frac{P \vee \neg P}{} \; Ax$$

$$\frac{P, \; Q}{P \wedge Q} \; \wedge\text{-}Intr \qquad \frac{P}{P \vee Q} \; \vee\text{-}Intr \qquad \frac{P \wedge Q}{P} \; \wedge\text{-}Elim$$

$$\frac{P \vee P}{P} \; Contr \qquad \frac{P, \; P \rightarrow Q}{Q} \; MP$$

$$\frac{P \vee Q}{Q \vee P} \; Symm \qquad \frac{\neg\neg P}{P} \; DN \qquad \frac{P}{\neg\neg P} \; DN$$

$$\begin{array}{c}[P]\\ \vdots \\ \dfrac{Q}{P \rightarrow Q}\end{array} \rightarrow\text{-}Intr \qquad \begin{array}{c}[P]\\ \vdots \\ \dfrac{\bot}{\neg P}\end{array} \neg\text{-}Intr$$

$$\frac{P \vee Q, \neg P \vee R}{Q \vee R} \; Cut \qquad \frac{P, \neg P}{\bot} \; Falso \qquad \frac{\bot}{X} \; EFQ$$

Remark

In the above, we have used one axiom which we know is always true, $P \vee P$. Note that $[P]$ is an assumption or a supposition in the proof process, which should be discarded down the deduction chain. Recall that this may be employed in the Fitch Style, and so it is true here as well. EFQ is Latin for "ex falso quadlibet". It means from a falsehood you can conclude or assert anything freely. This is why we used X—for any PL formula. This is fallacious of course. We are saying that if we allow a false assertion, we might as well conclude anything. This is seen in action when people present arguments filled with sophistry.

   Remember that we can have $\models F$ to say that F is a tautology. This is the semantic version, we have a syntax version of this as well.

Definition 6.6.3 *If we can prove a formula F by using only the inference rules above in PL, then we say F is a* theorem *of PL and we write it as* $\vdash F$.

Examples

Here we will prove formulas in PL that are theorems of PL.

1. $\vdash P \rightarrow (P \vee Q)$. Below we will make use of the assumption technique.

$$\cfrac{\cfrac{[P]}{(P \vee Q)} \vee\text{-Intr}}{P \rightarrow (P \vee Q)} \rightarrow\text{-Intr}$$

Note that $[P]$ has been discharged in the conclusion.

2. $\vdash \neg(P \wedge \neg P)$. Below start with Ax and note that we can use the equivalences found in Sect. 4.1.3.

$$\cfrac{\cfrac{\neg P \vee P}{\neg\neg(\neg P \vee P)} \text{Ax;DN}}{\neg(P \wedge \neg P)} \text{deMorgan}$$

Reflection 6.6.1 *See if using the rules you can prove to yourself that* $\vdash P \rightarrow \neg(\neg A \wedge B)$.

Examples

Let us try and prove theorems coming from a set of formulas Γ.

1. $\{P \wedge Q\} \vdash \neg(\neg P \vee \neg Q)$.

$$\cfrac{\cfrac{P \wedge Q}{\neg\neg(P \wedge Q)} \text{DN}}{\neg(\neg P \vee \neg Q)} \text{deMorgan}$$

2. $\{P \rightarrow Q\} \vdash (P \wedge R) \rightarrow Q$.

$$\cfrac{\cfrac{\cfrac{[P \wedge R]}{P} \wedge\text{-Elim} \qquad P \rightarrow Q}{Q} \text{MP}}{(P \wedge R) \rightarrow Q} \rightarrow\text{-Intr}$$

Reflection 6.6.2 *See if using the rules you can prove to yourself that* $\{P \rightarrow (Q \wedge R)\} \vdash P \rightarrow Q\}$.

6.7 Consistency, Completeness and Soundness

Consistency is a good property of a formal system to have because it ensures we can never produce a contradiction when we make deductions on it, since if we do succeed in such a goal, the formal system is not useful nor reliable. This is bad, it means it is internally inconsistent within its axioms and rules of deductions. Similarly we saw soundness before in Part I. There is no use having legally deduced theorem, if in the end, it is not true. So here let's make these concepts precise.

6.7.1 Particular

Here we start to see the effect of a set being on the left side of the turnstile $\vdash$. That set can be the empty set $\oslash$.

Definition 6.7.1 *We say that PL is* consistent *if there is no formula F of PL such that* $\vdash F$ *and* $\vdash \neg F$.

The above means we can never deduce both of them at the same time. In effect, we can never deduce an asserted formula F at the same time turn around and produce its negation $\neg F$ as well.

Below we have a different version of this. If we are able to produce either one of them but not both, then this is a good thing. Notice the "or" disjunction being used below

---

Definition 6.7.2 *We say that PL is* complete *if there is a formula F of PL such either* $\vdash F$ *or* $\vdash \neg F$.

Below we have the theorem of *soundness* for PL. We won't give a rigorous proof of this but will content ourselves with just an outline. It is because we need some of the intermediate lemmas and theorems to prove it. There is a lesson here though. In mathematics, as in logic, there could be a myriad of hoops to jump in order to get to the summit of a very useful theorem that has lots of good applications.

Theorem 6.7.1 (Particular Soundness Theorem) *If we have F a theorem of PL, i.e.,* $\vdash F$, *then F is a tautology of PL, i.e.,* $\models F$.

Proof We only give a sketch here.

The proof is using the principle of induction on the formulas of PL such that for each formula F being $\vdash F$ then we get $\models F$ too.

1. So here we can start with the base case where F is the result of the inference rule, such as the Ax rule. We know this is a tautology.
2. Then we proceed for the n as the I.H. that the soundness theorem above works for the first n formulas in the deduction. The aim now is to show that the theorem of soundness works for the $n + 1$ case too.

We refer to [2] for more details.   □

In the next section, we follow the same line of reasoning to prove an analogous result for a more general case.

Remark

It is a philosophical point if we notice that indeed $\emptyset \subseteq \Gamma$, but is this really a "particular" case? Perhaps, but it seems that it is more general in power than when $\Gamma \neq \emptyset$. Meaning, between $\vdash F$ vs $\Gamma \vdash F$, which one has a more extended result?

6.7.2 General

In this section we treat $\Gamma \neq \emptyset$. If we note our experience for example with plane or Euclidean Geometry, we observe that it starts off with postulates which are statements which are propositions of course. Then our teachers showed us how to derive new information about them, i.e., "theorems". The set of postulates can be said to form our Γ in our discussion here. It is the foundation of a "theory" which we alluded to in the previous section etc.

Below we have a general theorem for soundness. We won't prove this theorem because remember, our aim is to illustrate how we do proofs and we have been doing this now for most of the theorems we have encountered so far.

Theorem 6.7.2 (General Soundness Theorem) *If we have F as a theorem of Γ, i.e., $\Gamma \vdash F$, then $\Gamma \models F$.*

Remark

Fallacy Allert!

The general soundness theorem does not say that any F that can be deduced from Γ is automatically TRUE. It is only like that if Γ is TRUE, i.e., every statement in it is TRUE. Other than this, we are not guaranteed that F is TRUE. We can easily fall into this fallacy if we are not careful.

The converse of the above soundness theorems are great results to have but we need the theorem below to prove this.

Theorem 6.7.3 (The Deduction Theorem, Syntactic Version) $\Gamma \cup \{F\} \vdash G$ *iff* $\Gamma \vdash F \rightarrow G$.

Proof $\Longrightarrow$
The style is again using the induction on the length of the formula.
Base Case(s):

We have the case when in one step, we get G. Yet there are 3 ways this can happen, so this is proof by cases. It is sort of a case within a case type of reasoning.

1. G could be an axiom, that is one way a one step deduction can occur, so considering Γ alone we can have

$$\frac{\dfrac{G}{\neg F \vee G} \text{ $\vee$-Intro}}{F \rightarrow G} \text{ Impl}$$

∵ $F \rightarrow G$ from an axiom then $\vdash F \rightarrow G$, then of course we can assert that $\Gamma \vdash F \rightarrow G$.

2. Then either $G \in \Gamma$ or $G \in \{F\}$. If the first case, then we can assert G since it is in Γ, then we follow again the same situation as above. If the second case, then $G \equiv F$, we can assert $\vdash \neg G \vee G$ but this is the same as $\neg F \vee G \equiv F \rightarrow G$.

Inductive Step:
We now assume that the I.H. is true. So we assume that for the first n steps the conclusion holds, i.e., $\Gamma \vdash F \rightarrow G$. So we assume now that G is the conclusion from the $n + 1$ step. Then we need to show that again $\Gamma \vdash F \rightarrow G$ We have the following possibilities:

1. It is possible that G came from the above two cases above and so the same proofs apply.
2. It is possible that G came from the application of some inference rule. Our aim, therefore, is to show that whatever the inference rule used, we should still get the right side for this $(n + 1)$th step. We will show this for the case when MP has been used. So we have to show this for every rule but the argument idea is the same.

For the use of the MP rule, it would have looked something like this on the left side, i.e., on the $\Gamma \cup \{F\}$

$$
\begin{array}{c|l}
1 & \vdots \\[4pt]
2 & X \\[4pt]
3 & \vdots \\[4pt]
4 & X \to G \\[4pt]
5 & \vdots \\[4pt]
6 & G
\end{array}
$$

From above left-side situation and using the IH, we do have on the right, $\Gamma \vdash F \to X$ and $\Gamma \vdash F \to (X \to G)$ so

$$
\cfrac{\cfrac{\cfrac{\cfrac{\cfrac{F \to X, \quad F \to (X \to G)}{\neg F \vee X, \quad \neg F \vee (\neg X \vee G)}\ \text{Equiv}}{X \vee \neg F, \quad \neg X \vee (\neg F \vee G)}\ \text{Equiv}}{(\neg F \vee \neg F) \vee G}\ \text{Cut}}{\neg F \vee G}\ \text{Equiv}}{F \to G}\ \text{Equiv}
$$

$\therefore \Gamma \vdash F \to G$, which is required. We need to do the same for the other rules.

$\Longleftarrow$

Since $\Gamma \vdash F \to G$, we need now to consider $\Gamma \cup \{F\}$ and see if we can derive G from it. $\Gamma \vdash F \to G$ means $F \to G$ is deducible from Γ alone so if we add something to the left of Γ, the deduction is not affected. So

$$\Gamma \cup \{F\} \vdash F \to G$$

$$\Gamma \cup \{F\} \vdash F$$

$$\Gamma \cup \{F\} \vdash G \quad MP$$

which is what is required.   □

Reflection 6.7.1 *Can you extend the above proof for the other rules?*

Remark

Notice the length of this proof. It is quite a long process. Some proofs are like this and can span several pages too!

The deduction theorem has many applications, for example, it can shorten the proof of a theorem. More importantly, this has been used to prove the *Adequacy Theorem* below but not in a straightforward way. There are still intervening theorems and lemmas that must be shown in order to get the final result which we will state later. We come back now to the notion of *Consistency* and *Completeness* but this time with respect to Γ.

> **Definition 6.7.3** Γ *is* consistent *if for any formula F, we cannot have* $\Gamma \vdash F$ *and* $\Gamma \vdash \neg F$. *In effect, it is* consistent *iff* $\Gamma \nvdash \bot$

> **Definition 6.7.4** Γ *is* complete *iff for any formula or sentence F, it is the case that either* $\Gamma \vdash F$ *or* $\Gamma \vdash \neg F$.

The two notions are subtly confusing but we must try to have a clear head on this. Consistency says we can never prove a contradiction from Γ whereas completeness says given any formula ϕ, it is either we can have $\Gamma \vdash \phi$ or $\Gamma \vdash \neg\phi$. There are some Γ that are consistent but not complete so one does not imply the other but they are very good properties of a theory to have.

> **Theorem 6.7.4** (Adequacy Theorem) *If* $\Gamma \models F$ *then* $\Gamma \vdash F$.

Proof The proof of this requires in between lemmas and theorems involving consistency and satisfiability. Which we will not cover here but other logic books like that of [2] covers this. □

> **Theorem 6.7.5** (Completeness Theorem) *If* $\Gamma \models F$ *iff* $\Gamma \vdash F$.

> **Reflection 6.7.2** *Based on the theorems we have covered can you prove the Completeness Theorem?*

6.8 Resolution

Let's start with a scenario. Let us assume you and I both have ice-creams in our hand. Then I say to you, I will flip a coin, if it turns heads, I get you ice-cream. If it turns tails, you give me your ice-cream. Can you prove that I will always get your ice-cream? Logic with resolution can easily answer that.

Theorem 6.7.5 in a way is one such solution to the above problem. The above problem can be posed this way—given a set of propositions Γ can we derive F from there? We are of course asking is the case that $\Gamma \vdash F$? One way is to use a brute force approach to search for a derivation using the statements in Γ and the application of the inference rules to see if we get F. That is not viable, there can be lots of paths we need to try to get this and maybe it is not true that F is indeed derivable from Γ after all. Completeness Theorem to the rescue. So instead of answering the question directly, we can answer if $\Gamma \models F$. This is an if-and-only-if theorem and if the answer is no, then we have $\Gamma \nvDash F$, otherwise, we get $\Gamma \vdash F$. The procedure then is to create a (possibly big) truth table, do a conjunction on the statements in Γ and testing if the $\rightarrow$ in $\Gamma \rightarrow F$ yields all $\top$, i.e., valid.

However, this again is not always tenable that large truth table can hamper us down. So, in this section we learn a more efficient way of answering if $\Gamma \models F$ without resorting to truth tables. This method is called the *resolution method/calculus*.

6.8.1 Satisfiability and Consistency Again

Below are theorems we need to establish with the end goal of discussing the concept of resolution later.

Theorem 6.8.1 (The Deduction Theorem, Semantic Version) $\Gamma \cup \{F\} \models G$ *iff* $\Gamma \models F \rightarrow G$.

Proof We apply the Completeness Theorem on the Deduction Theorem above. □

We apply this theorem to prove the theorem shown below.

Theorem 6.8.2 $\Gamma \cup \{\neg F\}$ *is consistent ie,* $\Gamma \cup \{\neg F\} \nvdash \bot$ *iff* $\Gamma \nvdash F$

Proof We use the deduction theorem's syntactic version.
ook $\Longrightarrow$

$\Gamma \cup \{\neg F\} \nvdash \bot \Longrightarrow \Gamma \nvdash \neg F \rightarrow \bot \Longrightarrow \Gamma \nvdash F.$

$\Longleftarrow$

We use $\Gamma \nvdash F \Longrightarrow \Gamma \nvdash \neg F \rightarrow \bot \Longrightarrow \Gamma \cup \{\neg F\} \nvdash \bot$ □

Theorem 6.8.3 (Model Existence Theorem) *If Γ is consistent, i.e., $\Gamma \nvdash \neg \bot$, then Γ is satisfiable*

Proof Assume $\Gamma \nvdash \neg\bot \implies \Gamma\neg \models \bot \implies \Gamma \models \top$, thereby showing that Γ is satisfiable. □

We can state this in the contra-positive form.

Theorem 6.8.4 (Model Non-Existence Theorem) *If Γ is not satisfiable, then Γ is inconsistent.*

Reflection 6.8.1 *Can you show $\Gamma, P \vdash Q$ and $\Gamma, R \vdash Q \implies \Gamma, P \vee R \vdash Q$.*

6.8.2 The Normal Forms

Whatever form our formulas take, it can be transformed to the so-called *normal forms* which we define below. This form is useful in getting to our destination of finding if a set of formulas are satisfiable.

Definition 6.8.1 *A* literal *is an atomic formula or its negation.*
A formula F is in Conjunctive Normal Form (CNF) *iff it can equivalently be converted to this form $G_1 \wedge G_2 \wedge \ldots G_k$ such that each G_i is a disjunction of literals.*
A formula F is in Disjunctive Normal Form (DNF) *iff it can equivalently be converted to this form $G_1 \vee G_2 \vee \ldots G_k$ such that each G_i is a conjunction of literals.*

Examples

1. $q, \neg r, q \wedge \neg r, p \wedge (q \vee r) \wedge (t \vee \neg w)$ is in CNF
2. $q, \neg r, q \vee \neg r, p \vee (q \wedge r) \vee (t \wedge \neg w)$ is in DNF

Remark

Equivalences play a major role in this section. Recall too that we used the notation $\iff$ for equivalence. Do not be confused with this when we use $\equiv$, it is just for convenience. Consider $F \to G \equiv \neg F \vee G \equiv \neg(F \vee \neg G)$. We can do the same for the further definition of $\leftrightarrow$.

This example should convince us that whatever is the composition of F we can always make it look like a CNF or DNF.

Below we have some more equivalences of distribution:

1. $F \wedge (G \vee H) \equiv (F \wedge G) \vee (F \wedge H)$
2. $F \vee (G \wedge H) \equiv (F \vee G) \wedge (F \vee H)$.

Example

We will demonstrate how we can turn an arbitrary formula into either CNF or DNF. Let us do it for $F = P \rightarrow (Q \rightarrow (R \wedge S))$. Using equivalences we learned in Part I and in here, we have

$$\Longrightarrow$$
$$F = P \rightarrow (Q \rightarrow (R \wedge S))$$
$$F = P \rightarrow (\neg Q \vee (R \wedge S))$$
$$F = P \rightarrow ((\neg Q \vee R) \wedge (\neg Q \vee S))$$
$$F = \neg P \vee ((\neg Q \vee R) \wedge (\neg Q \vee S))$$
$$F = (\neg P \vee \neg Q \vee R) \wedge (\neg P \vee \neg Q \vee S)$$

As we can see, this is an example of a CNF. You can verify this by looking at the CNF definition.

Reflection 6.8.2 *Convert $F = (P \vee Q) \rightarrow (R \vee S)$ to CNF. Note, it might be that it looks like a DNF but the two are convertible to each other because of the deMorgan rule.*

6.8.3   The Method

The resolution method is a way of determining if $\Gamma \models F$ by refutation. We can get a hint of this method from Theorem 6.8.2. We want the form $\Gamma \cup \{\neg F\}$ side of this to be inconsistent or not satisfiable, so, no model, so F cannot be derived from Γ by the completeness theorem. It hangs on the concept what is called *resolvents* and the use of normal forms. This is the reason why we covered CNFs and DNFs. Because they are equivalent we will just make them conform to CNF.

This method is called the *resolution calculus*. We really need to prove that the method indeed produces the correct result and it is complete in the sense of Definition 6.7.4. However, in order to do this we need more machinery of theorems and lemmas to establish the trustworthiness of this calculus.

The topic is really meant for computer science majors and I do not presume my reader is necessarily wanting a deep dive into this. Besides we have limited space so we will just apply the method.

Definition 6.8.2 *Let C_1, C_2 be clauses. Assume that there exists an atomic proposition such that $a \in C_1$ its compliment be $\neg a \in C_2$. For example, if $a = p$ then $\neg a = \neg p$, if $a = \neg p$ then $\neg a = p$. Then the resolvent of C_1, C_2 is the clause that results $R = (C_1 - a) \cup (C_2 - \neg a)$*

Note that we are doing a set subtraction in constructing R.

Example

Let us go back again to $F = (\neg P \vee \neg Q \vee R) \wedge (\neg P \vee \neg Q \vee S)$, which we saw above. The clauses are those united by $\wedge$, and inside these clauses, they are united by $\vee$. Here we are assuming that P, Q, S, R are all atomic and if they are not, then the one which is not atomic should be driven down to their atomic components. Therefore we now have this:

- $C_1 = \{\neg P, \neg Q, R\}$.
- $C_2 = \{\neg P, \neg Q, S\}$.
- $R = \{\neg P, \neg Q, R, S\}$.
  This is the result of forming R, it is not empty nor reduced in elements because there is no complemented atomic formula to be used for set subtraction.

Example

Assume now that F is composed of the following clauses. Let us call this as Γ.

$$\Gamma = \{\{\neg p, q\}, \{\neg q, r\}, \{\neg q, \neg r, s\}, \{\neg s, t\}, p\}$$

We have five clauses in here. Let's call the following:
$C_1 = \{\neg p, q\}$
$C_2 = \{\neg q, r\}$
Then R of C_1, C_2 is $R = \{\neg p, r\}$. In effect, q cancels out with $\neg q$

Reflection 6.8.3 *Compute R for C_3 and C_4.*

Remark

It may be that after applying and applying the process of obtaining an R we get to the point when R becomes empty. Then we will follow the traditional way of designating this empty set in PL resolution calculus by the $\square$ symbol.

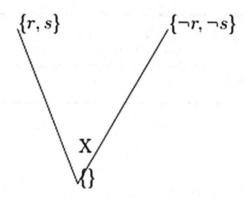

Fig. 6.5 Bad application of resolution

We formalize the continuous gathering of R by the definition below.

Definition 6.8.3 (The ρ Function) *Assume that Γ is a set of clauses. We define $\rho(\Gamma)$ by forming the union Γ with the R of any two clauses in Γ. Thus*

$$\rho(\Gamma) = \Gamma \cup \{R \mid R = (C_i - a) \cup (C_j - \neg a)\}$$

where $i \neq j$. We define also the following notations:

$$\rho^0(\Gamma) \equiv \Gamma$$

$$\rho^{n+1}(\Gamma) \equiv \rho(\rho^n(\Gamma))$$

for $n \geq 0$

$$\rho^* \equiv \bigcup_{n \geq 0} \rho^n(\Gamma)$$

Remark

Note: The cancelation that happens in *res* should be done one at a time, doing so provides a safe result. Look at the definition again and convince ourselves that is implied as well. See this illustration below using a tree graph. We have to do the resolution one at a time (Fig. 6.5).

We now present below without proof (as mentioned) the theorem that further underpins the resolution calculus

> **Theorem 6.8.5** (Resolution Theorem) Γ, *a set of clauses, is satisfiable iff* $\square \notin \rho^*(\Gamma)$. *Therefore also, Γ is unsatisfiable iff* $\square \in \rho^*(\Gamma)$

The Procedure

This is shown below in Fig. 6.6 but note, we need to visually check what we are canceling in this manual yet mechanical process.

Example

In the example below we will put the set of clauses at the top list. We read the resolution method from left to right. We start off with the first left most two clauses at the top. When we get R from this, we try to get a new R by comparing it to the next going to the right. We do this until all the clauses that are available have been gobbled up by the resolution method. Using the notion of ρ^* we would note that $\square$ is included in this big union. That means this Γ is unsatisfiable. This is found in Fig. 6.7.

> **Reflection 6.8.4** *Try to resolve these clauses* $\Gamma = \{\{a, b\}, \{\neg b, c, d\}\}$. *Did you get $\square$ in the process? What can you conclude?*

Demonstration

Assume we have the following information:

- $p =$ There is coronavirus in our area.
- $q =$ There will be a lock-down order by the governor.
- $r =$ We will quarantine at home.
- $s =$ We will communicate by social media.
- $t =$ We need Internet connection.

Furthermore, we have the following state of affairs:

- $p \rightarrow q$: If there is coronavirus in our area, then there will be a lock-down order by the governor.
- $q \rightarrow r$: If there is a lock-down order by the governor, then we will quarantine at home.
- $(q \wedge r) \rightarrow s$: If there is a lock-down order by the governor, and we will quarantine at home then we will communicate by social media.
- $s \rightarrow t$: If we communicate by social media, then we need Internet connection.
- p, is the case.
- Question: Do we need Internet connection?

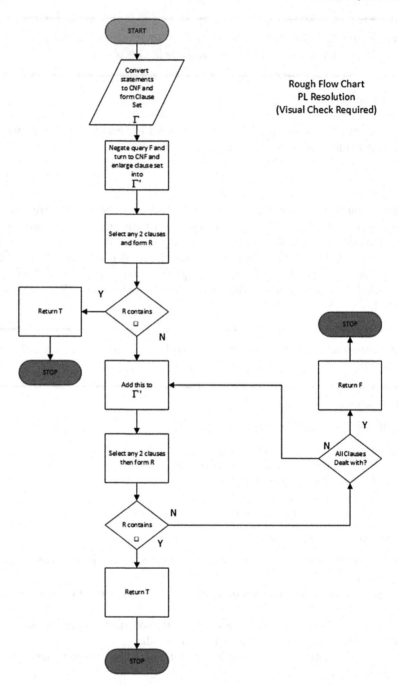

Fig. 6.6 Manual process of PL resolution

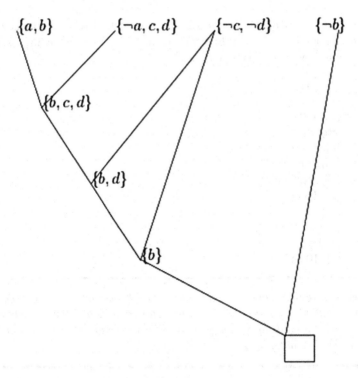

Fig. 6.7 Correct application—we get □ in the end

To proceed to answer the question, we turn all the statements above in CNF. Those statements become our Γ. Thus we have,

$$\Gamma = \{\{\neg p, q\}, \{\neg q, r\}, \{\neg q, \neg r, s\}, \{\neg s, t\}, p\}\cdot$$

Then we unionize this with the negation of t so we now have, and let us call this new set Γ'.

$$\Gamma' = \Gamma \cup \{\neg t\}$$

The right side is the familiar form found in Theorem 6.8.2. Γ' is the form found in the Resolution Theorem. So we proceed to use the resolution calculus found below but we need to re-arrange the clauses for neat visualization (Fig. 6.8).

We must note that it only looks like a tree because of the way it was arranged. However, it need not look like a tree! Below is another way of presenting the resolution. Steps 5 and 6 show that 3 has been re-used.

1. $C_1 = \{\neg p, q\}$
2. $C_2 = \{p\}$
3. $R_1 = \{q\}$, 1 and 2
4. $C_3 = \{\neg q, r\}$
5. $R_2 = \{r\}$, 3 and 4
6. $C_4 = \{\neg q, \neg r, s\}$
7. $R_3 = \{\neg q, s\}$, 5 and 6
8. $R_4 = \{s\}$, 3 and 7
9. $C_5 = \{\neg s, t\}$
10. $R_5 = \{t\}$ 8 and 9
11. $C_6 = \{\neg t\}$
12. $R_6 = \{\Box\}$ 10 and 11

Reflection 6.8.5 *Assume you and your friend have each an ice-cream. He invites you to play a game of heads or tails. If it lands heads, then he gets your ice-cream, if it lands tails, you give him your ice-cream. Prove that you will always wind up not having your ice-cream.*

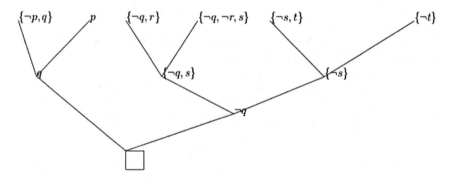

Fig. 6.8 Real life application—we get $\Box$ in the end

Fig. 6.9 J. A. Robinson

Remark

The resolution method was discovered by J. A. Robinson, a British-American philosopher turned mathematician. He is shown in Fig. 6.9.[2]

References

1. D. van Dalen, *Logic and Structure* (Springer, 2004)
2. R.E. Hodel, *An Introduction to Mathematical Logic* (Dover, 2013)
3. U. Schöning, *Logic for Computer Scientists* (Birkhäuser, Boston, MA, USA, 1989)

---

[2]Credits: David Monniaux, CC BY-SA 3.0, https://commons.wikimedia.org/wiki/File: John_Alan_Robinson_IMG_0493.jpg.

Chapter 7
Formal System for FOL

"It is not certain that everything is uncertain."

—Blaise Pascal

Abstract Like what we have done in the previous chapter, we will apply the same idea but this time to predicate logic otherwise known as *First-Order Logic* (FOL). Predicate logic is called FOL because the logic speaks about objects and their relationships with other objects. When we speak about something, we are predicating it, like in saying "John got infected by a virus". In general, FOL speaks about individuals. When we are predicating on a predicate itself, or we are speaking about sets of sets, then this is now the realm of *higher order logic* (HOL), which is beyond our scope.

7.1 FOL as a Formal System

We humans actually work and use FOL, it is part of our language. It is more expressive than PL in the sense that it gives us more information about individuals and so we can dig down deeply on a predicated subject. If we notice, from a language point of view, we can technically translate HOL to FOL and work in FOL but that becomes more complicated and complex to analyze. We won't bother with HOL.

Doing what we have done before previously, we apply the same ideas to draw theorems from the axioms and deduction rules found in FOL. The following ideas are helpful to recall before we proceed further.

1. Variables—refer to un-specified individuals, or un-named objects. For example, the variable may stand for humans, balls, integers, circles, etc. You name it, FOL can deal with it. Remember, it is the *domain* of discourse. We designate variables with traditional alphabets like x, y, z, u, v, w, etc. When we ran out of them, we can subscript them like x_1, w_2, etc.

© Springer Nature Switzerland AG 2021

L. P. Cruz, *Theoremus*,

https://doi.org/10.1007/978-3-030-68375-7_7

2. Constants—they stand for actual and named individual like "Bob", "the 8th ball", "320", "circle with r = 2 ins" etc. We designate them with the earlier part of the alphabet like a, b, c, etc., and the same as above, we can subscript them when we ran out of letters.
3. Functions—refer to mappings that designate one object in a domain to another object in another domain. Functions need to have an object source that it works on to correspond it to another object in another target domain. For example, we can say $f(x) = y$, meaning f operates on x that maps it to y. Since we are dealing with mathematics, we need functions to talk about and it would be less than adequate if we did not include this in our treatment. We did not have this in Sect. 4.2, but we now have it here.
4. Predicates—refer to what is happening to the object in question. So, things like "Bob is running", "the 8th ball dropped into pocket 1", "2 is rational", "circle A is crossed by a line", etc. As we saw previously we designate them with upper case letters like P, Q, S, T, etc.

In what follows, we may not provide proof for every theorem on FOL that we assert. It is because we have seen the method illustrated from the previous chapter, so we will usually skip some details.

7.2 FOL Syntax

7.2.1 Terms

The first concept we need to discuss is the notion of *terms*. The above objects we discussed in Sect. 7.1 except for predicates are these terms. We show in symbolic form how terms are constructed (Fig. 7.1).

For starters, x stands for any variable, so it implies too that we can use y, z, u, v, w etc. Then c stands for the constants, so symbolically we can use a, b, d, k, i, j, etc.

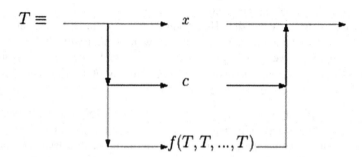

Fig. 7.1 Syntax diagram of terms

For functions f, g, h etc. As usual, it is implied that we can use sub-indexes when we run out of letters so x_1, c_1, f_1, etc.

Examples

Thus, we can have something like this: $f(g(x, h(y), z), y, a)$. Note in the above that T is recursive, so you can follow T back from the left and around, etc.

7.2.2 Formulas

The formulas of FOL are constructed from the following elements:

1. Terms
2. Predicates aka Relations

The predicates or relations are considered *atomic* or *primitive*, meaning, they cannot be broken down anymore, they are the given predicates of the FOL system we have. This is the same idea we had in PL. Predicates P, Q, S, T have "arities" or dimensions just like functions do. This means we can have $P(t_1, t_2)$, etc. Notice what the parameters as terms are inside P, and these terms are asserted to relate using the description provided by P. So we can actually have $R(Mary, Martha)$ with R standing for "sister of" thus, a way of saying "Mary is the sister of Martha". The arity or dimension of a predicate is always finite, just as they are in functions.

Figure 7.2 shows the syntax of FOL.

Remarks

1. The ts in the diagram stand for terms.
2. $t_2 = t_1$ and $P(t_1, t_2, \ldots, t_n)$ are called *atomic formulas*.
3. From the diagram we can then have formulas like $y = f(x)$ or $x = c$. This is a formula in FOL.
4. Having seen $\forall, \exists$ before, we are familiar with them and the syntax diagram implies to us that we can have a mixed quantification, i.e., we can have $\forall x_1 \exists x_2, x_3 P(x_1, x_2, x_3)$. This may be depicted in the syntax diagram below in Fig. 7.3.

Example

1. Let's turn to symbolic form the (wrong) assertion "All swans are white". How many predicates should we have here. We should have two, the being swan and the being white. So let S be for "swanness" and W for "whiteness". So we can translate this as $\forall x[S(x) \rightarrow W(x)]$.
2. Let's translate into symbolic FOL, the statement—For all natural number n, there is another m such that the product of m and n is also a natural number. So we need predicates, like N for being a natural number. We also need the predicate of being a product. Let's use a function $\times$ for this. So we have the following translation $\forall m \exists n[N(m) \wedge N(n) \wedge (k = \times(m, n)) \rightarrow N(k)]$.

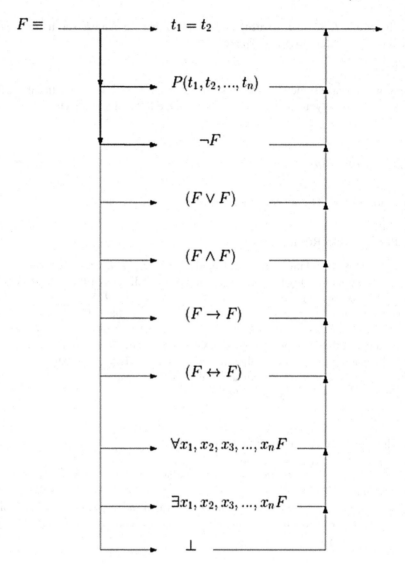

Fig. 7.2 FOL formula construction

3. We will translate the following

    a. All swans can fly: $\forall x[S(x) \rightarrow F(x)]$.
    b. Some birds do not fly: $\exists x[B(x) \rightarrow \neg F(x)]$.
    c. Hence, some birds are not swans: $\therefore, \exists x[B(x) \rightarrow \neg S(x)]$.

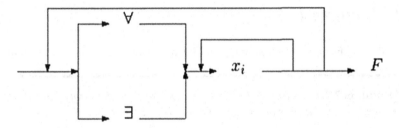

Fig. 7.3 FOL expansion of the quantification syntax

Reflection 7.2.1 *1. For the second example, explain why we can use a function for "product"? If we used a predicate P for this, would it be correct syntactically? Can you find an alternative way of expressing the one in the second example?*

2. Translate to symbolic form of: All men are mortals. Socrates is a man. Therefore, Socrates is mortal.

Lemma 7.2.1 *Let $\Phi(t)$ be a property and t be a term. If t is a variable or a constant or if $\Phi(t)$ such that $\Phi(t_1), \Phi(t_2), \ldots, \Phi(t_n) \Rightarrow \Phi(f(t_1, t_2, \ldots, t_n))$ holds, then $\Phi(t)$ holds for all terms.*

Lemma 7.2.2 *Let Φ be a property of formulas, assume the following:*

1. $\Phi(F)$ for F atomic.
2. F a formula, $\Phi(F) \Rightarrow \Phi(\neg F)$.
3. For F, G formulas, $\Phi(F), \Phi(G) \Rightarrow \Phi(F) \wedge \Phi(G), \Phi(F) \vee \Phi(G)$, $\Phi(F) \rightarrow \Phi(G), \Phi(F) \leftrightarrow \Phi(G)$.
4. $\Phi(F) \Rightarrow \Phi(\forall x_i F), \Phi(\exists x_i F)$ for every i

then $\Phi(F)$ for every formula F.

This is similar to the Theorem 6.3.1. The trick is to see that we are required to show that Φ follows, as a proof process.

7.2.3 Agreement

We will now discuss the notion of *free* or *bound* variables in relation to a formula F.

Definition 7.2.1 *Consider a formula $\forall x\, F$ or $\exists x\, F$. If x occurs anywhere inside F then we say that x is* bound *in F. Otherwise we say x* free

---

Example

1. $\forall xy(F(x, y, z)) \rightarrow \exists yz(G(x, y, z))$. In the first part left of $\rightarrow$, x, y are bound but z is free. To the right of $\rightarrow$, y, z are bound and x is free.
2. $\forall x, y A(x, y, f(z))$. Here x, y are bound but notice we have a term with the function $f(z)$, where variable z is free.

Remark

1. When a formula does not have free variables, we say the formula/statement is a *sentence*. Sentences are formulas where all variables are bound.
2. When we see a variable we may replace it with another term. For example in $\exists x A(x, y)$, y is free. If we substitute y with $f(x)$, then y is free because it is not mentioned in it. Similarly if we replace x in the predicate with y, then x holds free in that formula, but y is not.

Reflection 7.2.2 *Turn the following statement into symbolic form: Every prime number has one factor which is itself. Analyze all the variables in that formula and say if they are free or bound.*

7.2.4 Substitution

In Sect. 4.1.3, we saw some logical equivalences. Just like in PL, we may substitute one over the other when we see these forms. As always, we will need derivations too. We will see this in the future sections that will follow. Sometimes making use of these equivalences can make the derivations a lot simpler and economical.

There is another substitution that we need to cover aside from the above. This is the substitution of terms. We have to talk about this idea because it helps when we get to the question of the resolution process found in FOL. Recall the variables. These variables are placeholders for individual objects to stand in, they are place holder "names", so to speak.

So for example, take $\forall x P(x, y)$. Here we know x is bound but y is free. Since they are placeholders nothing is changed if we have $\forall v P(v, w)$. It says the same thing. However we may change the free variable with another term easily but we cannot change the bound term by another term without consequences of altering the sense of the statement.

Definition 7.2.2 (Notation for Term Substitution) *We write $[t/x]$ when we want to substitute t for x.*
$s[t/x]$ we write it this way when we want to change x by t in term s, and what results when the change happens.
$G[t/x]$ we write it this way when we want to change x by t in FOL formula G, and what results when the change happens.

Definition 7.2.3 (Rules for Term Substitution) *We assume that x is free in what follows below. Note we cannot carry out the replacement if this condition is not satisfied.*
$x[t/x] = t$, x cancels out.
$c[t/x] = c$, nothing to cancel out so it stays.
$y[t/x] = y$, the same as above.
$F(u, v, \ldots, w)[t/x] = F(u[t/x], v[t/x], \ldots, w[t/x])$, distribute the substitution.
$(t_1 = t_2)[t/x] \Rightarrow t_1[t/x] = t_2[t/x]$, same as above.

Example

1. Let us carry out $(x = y)[w/x]$. This becomes $x[w/x] = y[w/x] \Rightarrow w = y$.
2. Let us carry out $\forall w(x \cdot z = 1)[w/x]$. This becomes $\forall w(w \cdot z = 1)$.
3. Let us carry out $\forall w(x \cdot z = 1)[(w + y)/x]$. This becomes $\forall w((w + y) \cdot z = 1)$.

Reflection 7.2.3 *What do you think should result when we do $\bot[t/x]$?*

7.3 FOL Semantics

Just like we dealt with the semantics of PL we now do the same for FOL. However, in FOL it is not enough to have an interpretation function to determine truthfulness or falsity of a statement. FOL is more expressive than PL and FOL has a context in mind when statements are asserted. This is commonly known as a *domain of discourse*. Hence, we need a *structure* or an object with several components in that object to give a formula its proper meaning. We need a domain say U and an interpretation $[[\cdot]]^I$. These two have to go together always.

Definition 7.3.1 (A Model) *Let M be a structure with U as its domain and $[[\cdot]]^I$ its interpretation, then we will write $M = \langle U, I \rangle$.*
Then we define the way the interpretation works first for terms:

1. $[[c]]^I = c^I \in U$.
2. $[[f(t_1, t_2, t_3, \ldots, t_n)]]^I = f^I([[t_1]]^I, [[t_2]]^I), [[t_3]]^I, \ldots, [[t_n]]^I \in U$.

Let F and G be a sentences in FOL. Then we have the following interpretation of its syntax components:

1. $[[\bot]]^I = \bot$.
2. $[[t_1 = t_2]]^I = \top$ *iff* $[[t_1]]^I = [[t_2]]^I$, *else it is* $\bot$.
3. $[[R(t_1, t_2, t_3, \ldots, t_n)]]^I = \top$ *iff* $< [[t_1]]^I, [[t_2]]^I, [[t_3]]^I, \ldots, [[t_n]]^I > \in R^I$, $\bot$ *otherwise.*
4. $[[F \wedge G]]^I = \min([[F]]^I, [[G]]^I)$.
5. $[[F \vee G]]^I = \max([[F]]^I, [[G]]^I)$.
6. $[[F \rightarrow G]]^I = \bot$ *iff* $[[F]]^I = \top$ *and* $[[G]]^I = \bot$.
7. $[[\neg F]]^I = \neg [[F]]^I$.
8. $[[\forall x F]]^I = \top$ *iff* $[[F[a/x]]]^I = \top$ *for all* $a^I \in U$, *it is* $\bot$ *otherwise.*
9. $[[\exists x F]]^I = \top$ *iff* $[[F[a/x]]]^I = \top$ *for some* $a^I \in U$, *it is* $\bot$ *otherwise.*

Remark

We need to say something for the last two lines mentioning a^I. The object a^I in the real world is the individual that matches a in the syntax. This is because when apply I upon $[[F[a/x]]]^I$ we will eventually need to apply it to a, as in $[[a]]^I = a^I$. Another way of thinking about this is to see that under interpretation I, that a in our syntax is mapped to a said a^I.

Example

We will demonstrate how this model idea works. Let F be the statement
$\forall x \forall y \forall v \forall w[(E(x, y) \wedge E(v, w)) \rightarrow E(f(x, v), f(y, w))]$.

Can we find a structure that makes this statement true? Yes we can. If we set $U = \mathbb{N}$, the integers, and I be such that E is $=$ and $f(m, n) = m + n$, we will see that this structure makes this statement $\top$. Simply put, this comes from Euclidean Geometry's one of the many axioms, namely, if equals are added to equals, the wholes are equals too. For our purpose, when we speak of structure we normally mean a "mathematical structure". However, in computer science it is not necessarily like that, it can be any population of objects in the world.

Graphs and databases with their relationship rules are examples of structures and not just the numeric systems we gave above.

Definition 7.3.2 *Let F be a formula and $M = \langle U, I \rangle$. If $[[F]]I = \top$, then we say M is a* model *for F and we write $M \models F$.*

If it such a case that any model we can think of models F, then we write $\models F$ and say that F is valid. *Otherwise it is* invalid *and we write $\not\models F$.*

If there is at least one model for F, then F is satisfiable *else it is* unsatisfiable.

Further, let Γ be a set of FOL sentences, we write $M \models \Gamma$ iff for all elements $F \in \Gamma$ we have $M \models F$.

We also write $\Gamma \models F$ iff $M \models \Gamma$ implies $M \models F$, where $\Gamma \cup \{F\}$ are sentences.

Remarks

Recall the equivalences we mentioned in Part I, Sect. 4.2.4. They are useful to remember and can be used in the proving tasks. Similarly, we have the following Lemmas below. Try to prove them as an exercise.

Lemma 7.3.1 *Let F be any formula in FOL then*

1. *$\models \forall x F \leftrightarrow F$ provided x is not free in F*
2. *$\models \exists x F \leftrightarrow F$ provided x is not free in F.*

Reflection 7.3.1 *Can you try to prove the above to yourself? Hint: Apply the definition and note that x is not free in F, what does that mean?*

Lemma 7.3.2 *Let M_1 and M_2 having the same domain and interpretation then if F is an FOL statement, then $M_1 \models F \leftrightarrow M_2 \models F$.*

Proof We won't do a detailed proof on this but we start by the induction on a formula. So start with F is $s = t$ and apply the definition of $\models$ as you move on to the next form. $\qquad\square$

Theorem 7.3.1 *Let M be a model and let F, G be statements in FOL. If $M \models F \rightarrow G$ and $M \models F$ then $M \models G$.*

Proof From $M \models F \rightarrow G$, then by definition either $M \not\models F$ or $M \models G$. However, we were given that $M \models F$ so it cannot be the first in the previous OR statement, so it has to be the second, thus $M \models G$. $\qquad\square$

Reflection 7.3.2 *Try proving* $\mathcal{M} \models F \square G \Leftrightarrow \mathcal{M} \models F \blacksquare \mathcal{M} \models G$, *where* $\square$ *is* $\wedge, \vee, \rightarrow, \leftrightarrow$ *and* $\blacksquare$ *is "and", "or",* $\Rightarrow, \Leftrightarrow$, *respectively. In other words, put the connectives on the left and verify using the definition that the right will conform with the structure.*

Remark

The man responsible for this view of semantics is Alfred Tarski (1901–1983),[1] a Polish-American mathematician. We adopted this idea of "interpretation" as a function of being TRUE or FALSE when we were treating PL. This idea of semantics has now become a standard in many books and articles dealing many different types of logics, specially those found in computer science. See our references for more information.

We now turn to the corresponding Natural Deduction proof system for our FOL. See next section.

7.4 FOL Proof System

Since PL is subsumed in FOL, then all we need to do is add some rules that deal with quantification on top of PL. Furthermore, we recall the meaning of $\vdash$. For example, we retain the meaning of $\vdash F$ and $\Gamma \vdash F$ as before.

Definition 7.4.1 (FOL Quantification Rules) *Let* t *be any term and* x *be a variable.*

$$\frac{\forall x\, P(x)}{P(t)} \ \forall - Elim$$

$$\frac{P(t)}{\forall x\, P(x)} \ \forall - Intr$$

Provided in $\forall - Intr$, t *is only present in the derivation where it is introduced.*

$$\frac{P(t)}{\exists x\, P(x)} \ \exists - Intr$$

$$\frac{\exists x\, P(x)}{P(t)} \ \exists - Elim$$

Provided in $\exists - Elim$, t *is only present in the derivation where it is introduced. Note: We will follow the rules for equality, i.e.,* $=$, *in that it is reflexive, it commutes and is transitive.*

---

[1] https://en.wikipedia.org/wiki/Alfred_Tarski.

Remark

1. In the above t is usually c, a constant.
2. When it comes to equality of terms such as $s = t$, we are assuming implicitly that these are axioms in our system.
3. Using Leibnitz's principle, if two objects are identical, whatever can be predicated to one should be predicated to the other.

Definition 7.4.2 *If we can prove a formula F by using only the inference rules above in FOL, then we say F is a theorem of FOL and we write it as* ⊢ *F*.

Example

1. Let us try to prove $\forall x\, P(x) \wedge \forall x\, Q(x) \vdash \forall x (P(x) \wedge Q(x))$.

$$
\cfrac{
\cfrac{
\cfrac{\forall x\, P(x) \wedge \forall x\, Q(x)}{\forall x\, P(x) \qquad \forall x\, Q(x)} \wedge - Elim
}{
\cfrac{P(c) \qquad Q(c)}{P(c) \wedge Q(c)} \wedge - Intr
} \;\; \forall - Elim
}{
\forall x (P(x) \wedge Q(x))
} \;\; \forall - Intr
$$

2. Let us try to prove $\exists (F(x) \wedge G(x)), \forall x (G(x) \rightarrow H(x)) \vdash \exists x (F(x) \wedge H(x))$

$$
\cfrac{
\cfrac{
\cfrac{
\cfrac{\exists x (F(x) \wedge G(x)) \qquad \dfrac{\forall x (G(x) \rightarrow H(x))}{G(c) \rightarrow H(c)} \forall - Elim}{F(c) \wedge G(c) \qquad G(c) \rightarrow H(c)} \wedge - Elim
}{
\cfrac{F(c), G(c), \qquad G(c) \rightarrow H(c)}{H(c)} MP
}
}{
F(c) \wedge H(c))
} \rightarrow -Intr
}{
\exists x (F(x) \wedge H(x))
} \;\; \exists - Intr
$$

3. Let us try to prove $\forall P(x) \vdash \exists P(x)$

$$
\cfrac{
\cfrac{\forall x\, P(x)}{P(c)} \forall - Elim
}{
\exists P(x)
} \;\; \exists - Intr
$$

Reflection 7.4.1 *Consider the last example, is the reverse true? Try now deriving* ⊢ $\forall x\, P(x) \rightarrow \forall y\, P(y)$. *Hint, use the idea of assumption which we did in PL.*

7.4.1 Consistency, Completeness and Soundness

As with PL, the above ideas have corresponding versions in FOL. We will not deal
with the proofs in them in great detail because as we have said, our aim is to give the
reader only a flavor of how to do proofs applied to logic itself. We will only sketch
the proofs most of the time.

Below we diagram a possible way of looking how these concepts interconnect
with each other. We call this for now, "order of influence" of concepts. See Fig. 7.4.
It starts off from consistency and moves down to completeness.

Definition 7.4.3 *We say that FOL is* consistent *if there is no formula F of FOL
such that $\vdash F$ and $\vdash \neg F$. Assume that Γ is a set of formulas in FOL. Then we
say Γ is* consistent *iff there is no formula F in FOL such that we have $\Gamma \vdash F$
and $\Gamma \vdash \neg F$ at the same time.*

Definition 7.4.4 *We say that FOL is* complete *if there is a formula F of FOL
such either $\vdash F$ or $\vdash \neg F$. Assume that Γ is a set of formulas in FOL. Then we
say Γ is* complete *iff for every sentence F in FOL we have either $\Gamma \vdash F$ or
$\Gamma \vdash \neg F$.*

Theorem 7.4.1 (Particular Soundness Theorem) *If we have F a theorem of
FOL, i.e., $\vdash F$, then F is a tautology of FOL, i.e., $\models F$.*

Proof The proof of this is similar to that in PL except we employ the additional rules
and quantification formulas to show this result holds.   □

Theorem 7.4.2 (General Soundness Theorem) *Assume that $\Gamma \vdash F$, where F
is a sentence in FOL then $\Gamma \models F$.*

Proof Same style as above with the additional FOL rules being shown to comply. □

Theorem 7.4.3 *If Γ has a model then Γ is consistent.*

Fig. 7.4 Order of influence

Proof We prove by contradiction. Assume Γ has a model $\mathcal{M}$ and Γ is inconsistent. Then there is an F such that $\Gamma \vdash F$ and that $\Gamma \vdash \neg F$. By the soundness theorem then we have $\mathcal{M} \models \Gamma$ implies $\mathcal{M} \models F$ and $\mathcal{M} \models \neg F$. This means $\mathcal{M} \models F \wedge \neg F$ implies $\mathcal{M} \models \bot$. This is a contradiction.

$\therefore \Gamma$ is consistent.   □

Theorem 7.4.4 (The Deduction Theorem, Syntactic Version) *Let F, G be sentences in FOL, then $\Gamma \cup \{F\} \vdash G$ iff $\Gamma \vdash F \rightarrow G$.*

Proof Similar style as above.   □

Reflection 7.4.2 *Can you show why a model for $\forall x\, P \wedge \forall x\, Q$ cannot be a model for $\forall x(P \wedge Q)$? That is, show they are not equivalent.*

Theorem 7.4.5 (Completeness Theorem) *Let Γ be a set of sentences in FOL and F a sentence in FOL. Then $\Gamma \vdash F \Leftrightarrow \Gamma \models F$.*

Remark

We won't prove this result for FOL because in this work, it is not our purpose for the student to have a complete mastery of all aspects of logic We know this is already true in PL.

What is amazing about this result is that what can be proven syntactically is semantically true and vice versa. So there is correspondence between proof and models. This is a powerful property. This means that if we know that a model satisfies Γ also satisfies F, then we know there is a proof that will help us derive F from Γ. The mathematician who proved this was the American-Austrian man called Kurt Gödel.[2] See Fig. 7.5. This result though already astounding is yet not as clever as another theorem he is known for, it is called Gödel's Incompleteness Theorem. We won't get into this one but it is considered by mathematicians as earth shattering as Einstein's discovery of the Special Theory of Relativity.

---

[2]Unknown, http://www.arithmeum.uni-bonn.de/en/events/285, https://en.wikipedia.org/w/index.php?curid=43426215. From Wikipedia: The copyright of this work has expired in the European Union because it was published more than 70 years ago without a public claim of authorship (anonymous or pseudonymous), and no subsequent claim of authorship was made in the 70 years following its first publication.

Fig. 7.5 Kurt F. Gödel

7.5 Resolution

Like in PL, we can have the same method of analyzing if a FOL formula F is a consequence of a set of formulas Γ. If it is, then by the completeness theorem, the said formula is deducible from the set of formulas.

However, the process is rather more involved in FOL than in PL, for after all, as what we have seen, FOL is more expressive than PL. The idea works like this, if we have a set of formulas in FOL that can be made sentences, then as we know sentences are propositions. When this happens, since we now have propositions which we can deal with in PL, then we can just use the resolution process we have in PL and we are done. The target is to eventually get a CNF version of these formulas, those in Γ and F. For this reason, computer scientists are not prone to dismissing PL simply because it is less sophisticated than all of the logics out there. PL is very useful. Mathematicians often apply this philosophy of mapping a complicated object down to a known and properly understood object then solve the problem from the former by using the latter.

7.5.1 Rectified Form

> **Lemma 7.5.1** (Renaming Variables) *Let F be either of this form $\forall x\, G$ or $\exists x\, G$. Since x does not appear free in G, then we can replace it with another variable y so long as this does not occur free in G, as its equivalent.*

Proof This is obvious but, this can be proven by induction on the formula G.   □

Example

Let us consider $\forall x\, F(x, y) \wedge \exists y \exists x\, G(x, y)$. We cannot use y to replace anyone of the variables because y has an instance when it is free like in F. However we can do something with G since the variables there are bound. We can use $[y/w]$. Doing that for G, thus we get $[\forall x\, F(x, y) \wedge \exists w \exists x\, G(x, w)$. This new formula with renamed variables clearly says the same thing as the original one.

> **Definition 7.5.1** *A formula F and let F' is the result of the repeated application of Lemma 7.5.1 successively. Then we say F' is the rectified version of F if the quantifiers as a result are distinct and there are no variables (that refer to the replacement) that appear free in either F or F'.*

> **Lemma 7.5.2** *Given F a formula, then it is equivalent to a rectified F'. That is $F \equiv F'$.*

Remark

Basically, we are saying you can always rename the bounded variables in a formula F and we can continually apply renaming variables so that the resulting rectified formula is the same form as the original but with the proviso that the variables bounded by the quantifiers are named differently.

Some authors call the above process as *standardizing variables*.

Example

Let us consider this statement $\forall x(P(x) \rightarrow Q(x)) \wedge \exists x(R(x) \vee S(x)) \wedge \exists x\, R(x) \vee \exists z(T(z) \rightarrow S(z))$.

If we notice there is quite a repetition of x in that formula. We can apply rectification on the sub-formulas. There is a reason why this is useful to do which we will explain later. Let us try and make the variables controlled by the quantifiers as diverse as possible. Thus we can have the following result based on some preferences and get $\forall x(P(x) \to Q(x)) \land \exists y(R(y) \lor S(y)) \land \exists w R(w) \lor \exists z(T(z) \to S(z))$.

This is equivalent to the original. Yet, note that this is not the only possible way of rectifying the statement for you could have chosen your own preferred variable substitution so long as it does not violate the rules. Observe that the obvious equivalence of the original and new statement to each other.

Remark

We are going through the above exercise because we will need this technique in the resolution process. Eventually, this technique will help us get to our goal of transforming FOL statements that PL methods can handle.

7.5.2 Prenex Form

Definition 7.5.2 *We say a formula F is in* prenex form *iff*

1. *The formula is quantifier free, or*
2. *Let* $Q_i \in \{\forall, \exists\}$ *and* w_i *be a variable. Let also P be a formula with no quantifiers and F is of this form* $Q_1 w_1 Q_2 w_2 Q_3 w_3 \ldots P$, *where* $(i = 1, 2, 3, \ldots, n)$ *and* $w_i \neq w_j$.

Examples

Example	Prenex?
$R(x, y)$	Yes
$\forall x P(x) \to \exists y Q(x)$	No
$\forall x \exists y (P(x) \to Q(y))$	Yes
$\neg \exists x R(x, y)$	Yes

Reflection 7.5.1 *Consider the formula from the last section, namely,* $\forall x(P(x) \to Q(x)) \land \exists y(R(y) \lor S(y)) \land \exists w R(w) \lor \exists z(T(z) \to S(z))$. *Can you prove or disprove that this is equivalent to* $\forall x, \exists y \exists w \exists z[(P(x) \to Q(x)) \land (R(y) \lor S(y)) \land R(w) \lor (T(z) \to S(z))]$?

Remark

1. This technique helps us move toward making FOL formulas fit for PL resolution.
2. It is of special interest when P is in the prenex form and is in CNF, for this is what we have in Sect. 6.8.2.

7.5.3 Some Helpful Equivalences

In the following we assume that x does not occur free in F otherwise these won't work.

1. $F \wedge \exists x G(x) \leftrightarrow \exists x (F \wedge G(x))$.
2. $F \wedge \forall x G(x) \leftrightarrow \forall x (F \wedge G(x))$.
3. $F \vee \exists x G(x) \leftrightarrow \exists x (F \vee G(x))$.
4. $F \vee \forall x G(x) \leftrightarrow \forall x (F \vee G(x))$.

It is obvious that $Q x F \leftrightarrow F$, if F does not mention x. The quantifier $Q \in \{\forall, \exists\}$ is irrelevant in such situation.

In the following we no longer have necessarily the same assumption about F above.

In addition to the ones we saw in Sect. 4.2.4, we have the following negative equivalences.

1. $\forall x F \leftrightarrow \neg \exists x \neg F$.
2. $\exists x F \leftrightarrow \neg \forall x \neg F$.

We know the distribution of quantifiers work but not the reverse. Meaning, look at the last two below:

1. $\forall x (H \wedge G) \leftrightarrow \forall x H \wedge \forall G$.
2. $\exists x (H \vee G) \leftrightarrow \exists x H \vee \exists G$.
3. $\forall x (H \vee G) \not\leftrightarrow \forall x H \vee \forall G$.
4. $\exists x (H \wedge G) \not\leftrightarrow \exists x H \wedge \exists G$..

Remark

We can show why $\forall x (H \vee G) \not\leftrightarrow \forall x H \vee \forall G$ by an example. Let $H(x)$, to mean x an integer is odd, and $G(x)$, x an integer is even. So we have the sentence, for any integer x either it is odd or even, so $\forall x (H \vee G)$ is the corresponding symbolic form. But what does $\forall x H \vee \forall G$ mean? It says either all integers are odd or all integers are even, which is not the same as the original.

Reflection 7.5.2 *Can you find an example to show why* $\exists x (H \wedge G) \leftrightarrow \exists x H \wedge \exists G$ *is the case. Hence, the left side is not equivalent to the right side?*

7.5.4 Skolemizing

Let us consider $Q_1 w_1 Q_2 w_2 Q_3 w_3 \ldots P$ form. It would be nice if all of the Q_i are of one consistent quantifier form. We like to either make them all $\forall$ or all $\exists$. Also, there is something going for us in that we have seen $\forall$ and $\exists$ have a connection and we can translate one form to another. Remember we are still staying close to our goal which is the idea that we want to drop the FOL situation down to the PL situation and let the PL resolution method take over once we got the FOL formulas conforming to PL.

Skolemization is the technique of making the presence of $\exists$ in a formula disappear to another valid form. This helps in making all Q_i become $\forall$ making it more convenient to deal with. We said that a sentence is a formula that has only bound variables, so even if the quantifier is $\exists$, so long as it encloses the variables in that formula, then that formula is a sentence. We are quite fortunate, we have a situation we can leverage to our advantage in attaining our goal, i.e., turn the FOL into PL so the PL techniques can be applied and thus help us answer the question if one formula F, can be $\Gamma \models F$.

Essentially the method is to take that variable x bounded by $\exists$ and choose a function to name that object that is declared to exists.

This technique we are about to describe came from Thoralf Skolem[3] (1887–1963), a Norwegian mathematician. This is an example where mathematicians name a process or a theorem based on the one who either introduced it or discovered it. It is a way of honoring their peers for the helpful idea or technique their discovery contributes to the field (Fig. 7.6).

Definition 7.5.3 *Given a formula F, we say it is in* Skolem Normal Form (SNF) *iff*

1. *It is in conjunctive normal form, and*
2. *It is in prenex form, and*
3. *The quantifiers are all* $\forall$.

The above labels a formula having the above property. Note that all the requirements must be met simultaneously to be considered a formula to be in SNF.

---

[3] Unknown author, Public domain, via Wikimedia Commons.

Fig. 7.6 Thoralf Skolem

Examples

Example	SNF?
$R(x, y)$	Yes
$\forall x\, P(x) \vee \exists y\, Q(x)$	No, not in prenex
$\forall x \exists y\, P(x, y)$	No, prenex, but has $\exists$
$\neg \exists x\, R(x, y)$	No, prenex but has $\exists$
$\forall x \forall y (R(x, y) \wedge Q(x, y))$	Yes
$\forall x \forall y (R(x, y) \vee Q(x, y))$	Yes

Definition 7.5.4 (Skolemization) *Let F be a formula in FOL which is in prenex form, i.e., $F = Q_1 w_1 Q_2 w_2 Q_3 w_3 \ldots P$. We consider the following cases:*

1. *For every $\exists$ that appear after n $\forall$, replace it with a function of arrity n, i.e., $f(v_1, v_2, v_3, \ldots, v_n)$, where the v_i are the variables represented by the $\forall$.*
2. *For every $\exists$ that appear ahead of $\forall$ and is not preceded by any $\forall$ we simply replace it with a function of arity 1, i.e., $f(a_i)$ one for each $\exists$, where a_i being a constant.*

Then eliminate that $\exists$, thus deriving a new F^S from F.

The f we introduced would be a new f in addition to the existing ones in the domain. Such an f we call skolem function

---

Remark

1. Firstly, let us be mindful that the order of quantifiers matter and they are to be respected. So, swapping the position of the quantifiers changes the meaning of the statement. Consider $\forall x \exists y (x < y)$. This is always true in $\mathbb{N}$. Give us any $x \in \mathbb{N}$ we can give back a y that is greater than x. However, the reverse is not true for $\exists y \forall x (x < y)$. It says there exists a y, such that all x, x is less than y. For assume such a y, we can always get a number x such that it falsifies $<$.
2. Using the above technique we get to have an SNF from a given F.

Example

In the following examples we will do skolemization with the intent also to have SNF.

1. Skolemize: Every mathematician drinks at least one coffee a day. We have $F = \forall x (Math(x) \rightarrow \exists y (Coffee(y) \wedge Drinks(x, y)))$. Turn F to a disjunction so we get CNF, so $\forall x (\neg Math(x) \vee \exists y (Coffee(y) \wedge Drinks(x, y)))$. Now we replace $\exists y$ with a function f, resulting $F^S = \forall x (\neg Math(x) \vee (Coffee(f(x)) \wedge Drinks(x, y)))$. Here we are saying f is the function that given an x, it will find for us the coffee that mathematician x drinks.
2. Skolemize: There exists a being such that all other beings are cared for by this being. $F = \exists x \forall y (Being(x) \wedge Being(y) \wedge CaredFor(y, x) \wedge (x \neq y))$. Already in prenex we just eliminate $\exists$, so identify x as GOD. We get $F^S = \forall y (Being(GOD) \wedge Being(y) \wedge CaredFor(y, GOD) \wedge (GOD \neq y))$.
3. Skolemize: $\forall x \exists y \forall z \exists w P(x, y, z, w)$. This becomes $\forall x \forall z P(x, f(x), z, g(x, z))$.

Theorem 7.5.1 *Given a formula F of FOL, then F is satisfiable iff its F^S is satisfiable.*

Proof We won't supply a proof in great detail but we will just go through it in outline form.

The proof is by induction once we get the F into prenex form. We prove that when F is satisfiable, i.e., has a model, then show F^S has a model too. We do this for the following cases:

1. The case when $\exists$ is at the front.
2. The case when $\exists$ is at the middle or last.

In both cases, we assume F has a model then prove that F^S has as well and vice versa. $\square$

Remark

1. We hope the reader notices the technique used in the above, that is, proving by cases.
2. Note that the above does not mean they are equivalent in every way, or in every model. It is only true if either one has a model. If there is no model for either one of them, the other has none either.

7.5.5 Unification

We recall that the idea of resolution came from J. A. Robinson, someone we met in the previous chapter.

Our goal is to have substitutions that will make the formulas term wise the same. We want this to happen after we have Skolemized the formula. For when we do, we are now in the realm of PL since they are sentences just like PL statements.

We have to go back and recall the idea of a literal but this time in the field of FOL.

Definition 7.5.5 *Any relation in FOL or its negation is called* literal *in FOL.*

Examples

1. $R(x, y)$, $Mother(mary)$, $R(f(a), y)$, $R(x, c)$, $\neg Q(a, f(y))$ are examples of literals.
2. c, $f(x)$, x, $f(y)$ are not examples of literals because they are not relations!

Definition 7.5.6 *Let $\mathcal{L}_1$ and $\mathcal{L}_2$ be literals, if there is a substitution of terms in $\mathcal{L}_1$ and $\mathcal{L}_2$ such that the effect of the substitution makes $\mathcal{L}_1 = \mathcal{L}_2$, then we say that $\mathcal{L}_1$ and $\mathcal{L}_2$ are* unifiable.

Call also such substitution s, since $\mathcal{L}_1[s] = \mathcal{L}_2[s]$, then we say s is a unifier *of $\mathcal{L}_1$ and $\mathcal{L}_2$.*

Examples

1. Let $\mathcal{L}_1 = P(a)$ and let $\mathcal{L}_2 = \forall x P(x)$. We can let $s = a/x$. Applying them to each literal we get $\mathcal{L}_1[s] = P(a)[a/x] = P(a)$, $\mathcal{L}_2[s] = \forall x P(x)[a/x] = P(a)$ so the two become the same.
2. Assume we have a set of literals $\mathcal{L} = \{P(x, g(a)), P(b, y)\}$, if we try $s = [b/x, g(a)/y]$, we transform $\mathcal{L} = \{P(b, g(a))\}$.
3. Assume we have a set of literals $\mathcal{L} = \{P(w.w, z), P(w, z, c)\}$. There is no possible unifier for this set.
4. Try it for $\mathcal{L} = \{P(x, f(x)), P(y, y)\}$. There is no possible unifier for this set.

Remark

1. If we observe it is not possible to unify differing predicates like $\mathcal{L} = \{R(x, y), S(x, y)\}$.
2. Different corresponding functional symbols make it impossible to unify like $\mathcal{L} = \{R(f(x), y), S(g(x), y)\}$.
3. When a variable is found in a corresponding spot but in a function make it impossible to unify. E.g., $\mathcal{L} = \{Q(x), Q(h(x))\}$.

Reflection 7.5.3 *Try unifying* $\mathcal{L} = \{Q(a, f(x), Q(y, f(w)))\}$.

7.5.6 The Procedure

Having done the preliminary work we are now ready to put everything into a process. There are many variations on the step-by-step process depending on the author presenting the technique, so be aware of this. You may see ordering differences in the in-between steps.

In what follows, just like in PL, we should first negate the target formula we are testing to see if it is derivable from the set Γ. We throw this into the mix and then we do the process of cleanly making the FOL formulas conform to PL manipulation. In Fig. 7.7, the first step is to push those negations further inside. Then the process of dealing with the quantifiers follows. Further, we recall that in the application of the process, whenever we get a proposition and its negation together, they cancel out or disappear from further treatment. When we arrive at an empty set, then we conclude we got a contradiction from the negation of the statement in question. Thus we affirm that the statement does follow from the set. However, if we are left with a non empty set to further resolve but cannot do so anymore then we conclude, the said formula does not follow from the set Γ.

The reader can also refer to [1] as an alternative procedure.

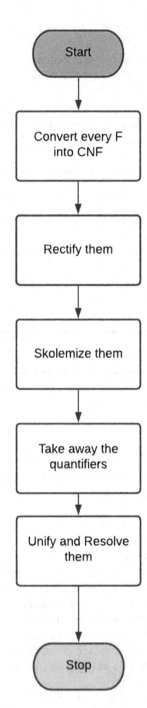

Fig. 7.7 FOL resolution process

Example

Let us confirm if the following syllogism is valid. Let S, T stand for *Student Of* and *Taught By* respectively. Assume we have the following assertions: $\exists x \forall y S(x, y)$, $\forall x \forall y [S(x, y) \rightarrow T(x, y)]$. Can we conclude $\exists x \forall y T(x, y)$?

We can look at it this way:

$$\frac{\forall x \forall y [S(x, y) \rightarrow T(x, y)], \quad \exists x \forall y S(x, y)}{\exists x \forall y T(x, y)}$$

1. We negate the conclusion, $\neg \exists x \forall y T(x, y) \Longleftrightarrow \forall x [\neg \forall y T(x, y)] \Longleftrightarrow \forall x \exists y \neg T(x, y)$. This in the first place is already in CNF and we do not need to rectify as we have a simple variables. Notice how we used the equivalences of negated $\exists$ and $\forall$. We skolemize to $\forall x \neg T(x, f(x))$. We remove the quantifiers and we get $\{\neg T(x, f(x))\}$.
2. We deal now with $\forall x \forall y [S(x, y) \rightarrow T(x, y)] \Longleftrightarrow \forall x \forall y [\neg S(x, y) \vee T(x, y)]$. This is in CNF now and no rectification required because we do not have $\exists$. We are left with $\neg S(x, y) \vee T(x, y)$, thus $\{S(x, y), T(x, y)\}$.
3. Coming now to $\exists x \forall y S(x, y) \Longleftrightarrow \forall y S(a, y)$ giving us $\{S(a, y)\}$.

We are left with three clauses and we can now draw the diagram for the refutation process see Fig. 7.8. Notice that we have to do the unification process by doing a substitution process so we are able to cancel the terms that are compliments of each other. In this case we are left with the $\square$ empty set and so this shows that the conclusion does follow from the premises.

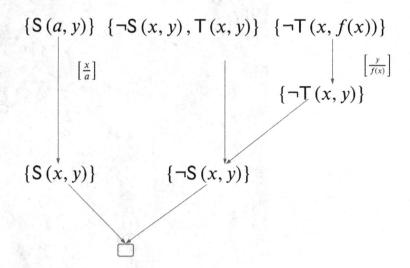

Fig. 7.8 FOL resolution example

Reflection 7.5.4 *Using what we have learned validate if the conclusion below is correct:*

1. *Santa Claus knows if you have been naughty or nice.*
2. *If you have been naughty, Santa will pass you by.*
3. *If you have been nice, Santa will give you a lovely toy.*
4. *Ivy has been nice.*
5. *Conclusion: Ivy will receive a nice toy from Santa.*

Remark

1. We went through this lengthy presentation to give the reader an idea that the topic of resolution has great and useful application in real life. The most significant one is in the area of "automated theorem proving" which is a sub-topic of Artificial Intelligence in Computer Science. Thus, we can ask the computer to do the resolution for us. Rather than us humans figuring out what to conclude from a bunch

Fig. 7.9 Hilary Putnam

of statements, we can have the computer do that for us. For those of us who have an advanced level of curiosity, such as those coming from computer science, the reader may refer to several advanced referenced works like that of [2, 3] and etc. It is believed that after getting this far in this book, the reader should be ready to tackle the more intricate aspects of our subject found in the works we have listed as references of this book.

2. We said in Part I that the resolution technique for validating if a formula can be concluded from a set of premises is attributed to J. A. Robinson. Actually there were two American mathematicians before Robinson who solved this problem. They were Martin Davis[4] and Hilary Putnam[5] (Fig. 7.9) in 1960. However, theirs were more of a search algorithm that can be computationally expensive. Robinson provided a more efficient method which is syntactic in nature, what we have tried to do here.

References

1. J. Barwise, J. Echemendy, *Language, Proof and Logic* (Seven Bridges Press, 1999).
2. S. Hedman, *A First Course in Logic: An Introduction to Model Theory, Proof Theory, Computability, and Complexity (Oxford Texts in Logic)* (Oxford, New York, NY, USA, 2004)
3. A. Galton, *Logic for Information Technology* (Wiley, New York, NY, USA, 1990)

---

[4]https://en.wikipedia.org/wiki/Martin_Davis_(mathematician).

[5]Unknown author, CC BY-SA 2.5 https://creativecommons.org/licenses/by-sa/2.5, via Wikimedia Commons.

Chapter 8
Doing the Math

I have no particular talent. I am only inquisitive.

—Albert Einstein

Abstract Previously we alluded to the idea of theories. Here we get to define what they are. We want to be concrete; it is not interesting to be confined to the abstract. We want to show that the previous lessons we learned have useful applications, to mathematics itself. We like to give the reader the truth of how theorems come about, that they do not come by easily, that it is at most times a foggy arduous affair without a clear process.

8.1 First-Order Theories

We consider again FOL. If we recall we kept on mentioning Γ. We often see this form—$\Gamma \vdash F$. Γ we know is a set of premise statements and F we said can be derived from these premises by use of the deduction rules. Therefore F is one of the theorems that can be pulled out from Γ.

This is what is happening in mathematical theories which you have encountered in your studies. From your elementary days, to your high school days and even probably now, your university days, you are encountering these said theories. For example, if you ever took Plane Geometry (Euclidean Geometry) we can view Γ as the postulates in geometry and look F as one of the theorems like "the total internal angles in a triangle is 180°". When we talk about mathematical theories we mean an area of study like Number Theory, Set Theory, Group Theory, Arithmetic, Category Theory[1] etc. So we can see Γ as the axioms of a theory. Don't be shocked if I tell you there are a myriad of broad theories in Mathematics!

---

[1]The reader can check the meaning of these theories using the Internet.

© Springer Nature Switzerland AG 2021
L. P. Cruz, *Theoremus*,
https://doi.org/10.1007/978-3-030-68375-7_8

Sometimes, mathematicians may refer to these theories with their attending algorithms as "mathematical machinery" or "mathematical infrastructure". Mathematical theories are much more solid and more stronger than "theories" formulated in other sciences like Biology, Medicine, etc. because we have seen that a theory in those areas can be re-stated after more "evidence" surface on a subject. Mathematicians don't accept assertions unless there is a proof that can be demonstrated deductively that the said assertion follows from the premises logically. It is culturally acceptable among mathematicians to be skeptical and doubtful. Mathematics does not move forward using inductive logic but by deductive logic. These mathematical structures are often referred to as *first-order theories* because they come from FOL and uses FOL as the lens through which they are examined.

Should the reader like to go deeper into these parts of mathematics like computing, then many ideas get cleared up when the reader grasps this notion of first-order theories. Most of the time this idea is taken for granted. We will walk through the composition of first-order theories below.

8.1.1 Definition of Theories

A *first-order theory* is composed of the following:

1. The language of FOL.
2. The symbols found in FOL, namely, the predicates along with $=$, operations/functions and constants. These are called *proper symbols* which we designate by $\mathcal{P}$
3. The logical axioms of FOL, which here we called Λ.
4. A set of *sentences/statements*, i.e., set of formulas with bound variables which here we designated as Γ. These are called *proper axioms/non-logical axioms*. Some like DeLong [1], refer to these as *initial formulas*.
5. The rule of inference is as before, that of FOL.

We must note that some authors define a theory this way: T is said to be a theory iff $T \models \sigma$ where σ is a sentence, implies that $\sigma \in T$. This manner of speaking is called being "closed" under validity. The reader might be confused as there are authors who define it in another way, that of being closed under derivability, meaning T is a theory iff $T \vdash \sigma$ implies $\sigma \in T$. This is Gödel's completeness theorem (we have studied previously) in action.

In computer science, there is an idea called a *knowledge base*. A knowledge base is a set of statements from which we can deduce new information from what has been given. What happens here is that we encode assertions into a language similar in pattern-like FOL known as *knowledge representation*. They are of course more close to our natural language and based on that we can check and inquire it whether or not an assertion we thought of can be drawn from the said knowledge base. Essentially this is what happens when we deal with a programming language like Prolog. In this case, those statements that a user throws into the knowledge base can be thought of as proper axioms in the sense they are on top of FOL axioms as they are our statements, a level up from the accepted FOL axioms.

Some authors like Hamilton [2] say that a theory then "extends" FOL because of the additional proper axioms. The word "extend" is subject to much interpretations, and how I understand it is that these proper axioms have a way of shaping the resulting system, i.e., they influence the final properties of the resulting system. For example, proper axioms can limit the models that will satisfy its statements. If we consider the statement $\exists x P(x) \to \forall x P(x)$ to be a proper axiom, then the only model that can satisfy this is a model with only one element in its domain (see DeLong [1]).

On the other hand the proper axiom
$$\exists x \exists y ((P(x) \land P(y)) \land (Q(x) \leftrightarrow \neg Q(y))) \to \forall x P(x)$$
implies that models with domain population of two elements are the only ones that can satisfy this axiom [1]. Hence, proper axioms have a way of dictating how the system eventually pans out. In short, proper axioms have a way of pulling or pushing a theory in a certain direction.

In studying first-order theories, it should be remembered that all concepts such as deducibility, models, etc. are still as before.

Authors like Margaris [3] make things explicit. Margaris likes to precisely designate with the notation $\vdash_T$ to inform the reader that the derivation comes from considering the proper axioms of T. You will see this type of notation in the literature.

Once can note that what we have done was to abstractly generalize these theories. Hence, we may say the object of our study are mathematical theories and we reasoned about them using logic. Thus, our study may also be called *mathematical logic* as well.

8.1.2 Some Examples

The inspiration for these is found in Margalis' work [3]. Below is an example of a mathematical theory expressed in FOL.

Theory of Groups

1. $\mathcal{P} := \{=, \cdot, 1\}$.
2. Proper Axioms

    a. $\mathcal{G}1. \forall a \forall b \forall c ((a \cdot b) \cdot c = a \cdot (b \cdot c))$.
    b. $\mathcal{G}2. \forall a (a \cdot 1 = a)$.
    c. $\mathcal{G}3. \forall a \exists b (a \cdot y = 1)$.

The above is an example of a mathematical theory where massive theorems have been proved even being applied to physics of electrons. Now, you might have not heard of *category theory*, but for interest perhaps you should briefly read into this. It is interesting to note that in relation to category theory, the above axioms are also its axioms if we make the variables x, y, z refer to functions. I know this might go above the student's head but I only mention it so that the reader is made aware that in mathematics, the abstraction is such that it can be applied to some patterns that make them theoretically equivalent.

Here we have an example where one theory, the theory of groups is subsumed in the theory of rings.

Theory of Rings

1. $\mathcal{P} := \{=, +, \cdot, 0\}$ plus the symbols of a group.
2. Proper Axioms

   a. $\mathcal{R}0$. A ring is a group, plus axioms below.
   b. $\mathcal{R}1$. $\forall a(a + 0 = a)$.
   c. $\mathcal{R}2$. $\forall a \exists b(a + b = 0)$.
   d. $\mathcal{R}3$. $\forall a \forall b(a + b = b + a)$.
   e. $\mathcal{R}4$. $\forall a \forall b \forall c((a + b) + c = a + (b + c))$.
   f. $\mathcal{R}5$. $\forall a \forall b \forall c(a \cdot (b + c) = a \cdot b + a \cdot c)$.
   g. $\mathcal{R}6$. $\forall a \forall b \forall c((b + c) \cdot a = b \cdot a + c \cdot a)$.

Remark

Thus far, we can see from the above how the whole infrastructure of a mathematical theory can be built and grow and move from a starting proper symbols and proper axioms. Since a group $\subset$ a ring, then any theorem for a group becomes a theorem in a ring that encloses it. Using these along with logic's deduction rules, we can start deriving new F. We then add this new F, since it is a theorem, to our Γ and so get more new Fs.

If you have time you might look at [3] to see this action of producing new theorems based on the theory's axioms.

FOL

First-Order Logic Itself

FOL itself can be viewed as a theory! This is somewhat not immediately obvious but we can think of it as a theory with the proper axioms being an empty set, i.e., no proper axioms just the proper symbols and the logical axioms and no more. This fact is quite significant in clarifying some oversimplifications and bad impressions we may have of **FOL**. When **FOL** is considered a theory it is called *Predicate Calculus* as we said before.

Another common misconception among computing people, this time, is that **FOL** is all there is and nothing more matters. They think if it is not **FOL**, it is not good enough. However, there are other logical systems that are quite useful and need not be as expressive as **FOL**. One should not limit his/her study to **FOL** for it has fragments or weakened subsets that are of practical use in computing too.

Do you remember what we did in Chap. 7? We wanted to drop FOL statements down to PL so that when we get there we can use the techniques for resolution that are available in that world! We can picture it this way:

$$F^{FOL} \rightarrow F^{PL} \rightarrow Res$$

Formulas in FOL are transformed to formulas in PL and when we get there we allow the methods in resolution to take over.

This is familiar idea employed by mathematicians. They ask what is it we know? What is it we don't know? Could we find a mapping such that we can take what we don't know to what we do know?

8.2 The Production of Theorems

When you go through a mathematics textbook and find some theorems which are presented to you, I am sure you have been amazed and must have mumbled—*How in the world did they know that?* What you mean is—how or where did they get that? This same amazement comes to us when we find exercises in the textbook challenging us to prove a statement. Mathematicians seem to hand these statements to us on a silver platter. They come to us ready for us to attack and prove. It seems like magic, like they plucked these exercises from the air. The process seems to be so neat and clean, so we thought. It only seems that way.

Actually, it is a lot messier than we think. It is really a foggy affair. In reality, there is a bit of chaos that goes on before they come up with these finely constructed exercise statements. Mathematicians stumble on theorems using several interconnected processes but they start with a suspicion or a conjecture. They formulate these conjectures using the following ideas:

- They rely on their intuition and similar experience.
- They imagine on what might be possible and use their power of observation.
- They might consider symmetry and beauty found in nature guessing that it might be the same.
- They might perform mental experiments doing pen and paper work.

However, before we discuss in sincerity how serious mathematics is done we need to talk about pseudo proofs first.

8.2.1 Crank Proofs

We will first give an example of a crank proof before we define what it is. We want to discuss this topic because we do not like to be categorized as a *crank*. You can guess, it is not a badge of honor to be called one.

UnTheorem 8.2.1 $2 = 1$, *we have proof of this.*

Proof Let $x = y$.

Using algebraic rules we have
$$\Rightarrow xy = y^2$$
$$\Rightarrow y^2 = xy$$
$$\Rightarrow y^2 + x^2 = xy + y^2$$
$$\Rightarrow 2y^2 = y^2 + xy$$
$$\Rightarrow 2y^2 - 2xy = y^2 + xy - 2xy$$
$$\Rightarrow 2y^2 - 2xy = y^2 - 2xy$$
$$\Rightarrow 2(y^2 - xy) = y^2 - xy$$

$$\Rightarrow \quad 2 = \frac{\cancel{y^2 - xy}}{\cancel{(y^2 - xy)}} = 1$$

$\Rightarrow 2 = 1$ $\qquad\qquad\qquad\qquad\qquad\qquad\qquad\qquad\qquad\qquad\qquad$ □

If you are feeling bewildered, you should! This cannot be correct, but if you look at the steps, all of them looked legal!

This is what we call a *crank proof*. You already came across some fallacies in Part I, here is another one. There is an air of seriousness on this one, right?

Somewhere in the steps there was a non-obvious illegal argument, and a person who engages in these fallacies often is a *crank*. The art of doing this is called *crankery*.

Most crankeries happen in the field of mathematics but now, it happens more in physics. At times, the crank proof would come from an amateur but ambitious student of mathematics. Some of them exercise skepticism for existing proofs or for the needless complexity of how to solve open mathematical problems. They suspect the problem is less complex than what is presently believed.

This is good, for people should not accept something as a fact unless they have a proof of what is being wagered. We should be like this, and not be gullible. However, there is a limit to this incredulity. An example is the attempted proof we showed proving we can make $2 = 1$. Right away, using what we know even in high school mathematics, we should have alarm bells ringing at such a thought.

There are myriads of open problems in mathematics and theoretical physics. Of course, you could be hailed as a hero if you provide a proof for solving these problems. It is indeed tempting to a student to attempt to solve one or two of them. Some of these open problems have monetary and honorific awards attached to them. Not only could they win a prize and enjoy this income, but it could spell a glorious career ahead if they manage to succeed. Most of these people are not really malevolent cranks, rather, they could just be innocent people who commit a blunder in their attempted proof. Committing a fallacy, we said can affect anyone, even long careered mathematicians. All of us are frail human beings susceptible to careless mistakes.

Mathematics is the only discipline whose data is abstract but objective. It is not like any of the sciences whose theory today could be adjusted tomorrow. Scientists outside the discipline marvel at its unique effectiveness [4]. The tradition of requiring proof from sound logical deduction is a value that a student of mathematics should propound as an inherent principle. Personality wise, I have found mathematicians to be less offended when it came to being questioned. They are not shy when it came to challenges, they welcome the chance to explain. Proof in mathematics is a community activity. Recall what we have said, a proof is meant to convince the legality of the arguments found in it.

Questioning even long held beliefs produces new knowledge to be found. We will see in the next section that theorems are born this way. For example, it is the exercise of conjecturing that propels research and stumbles on new discoveries. It is good to know that this art of questioning is still practiced today; one example of this is the use of mathematics to give a skeptical angle to Darwin's Theory of Evolution [5].

Reflection 8.2.1 *In the above steps, determine the illegal move. Hint: what are x and y?*

8.2.2 Real Serious Proofs

One way to learn how a theorem is discovered is by reading a biographical book about a famous mathematician. There are many of these biographical works. An example is Cedric Velani's account of how he, along with Clement Mouhot, formulated their theorem of 2009 [6]. We won't restate their theorem here since it is too advanced for us to deal with. We will just say for interest mention that it has something to do with Fourier transform and the so-called Landau Damping in theoretical physics. Villani described their proof by construction.

Vilani[2] (shown in Fig. 8.1) mentioned the torturous journey he and his collaborator went through starting in 2008 and finishing in 2009. So a total of 1 year to clinch it. In his account Villani compared it to a quest not sure if they will stumble on a result worth sharing or announcing. He would confess that the journey was not a straight line, it can hit a snag combined with some disillusionment too. We just mentioned previously about fallacies or crankeries; it is interesting to note that Villani described his feelings of doubt fearing that the reviewers of their proof might find an error they might have overlooked somewhere or that positive assessments made by reviewers get negated by another set of mathematicians.

There is a lot of trial and error that happens before a statement gets the status of a theorem. There is plenty of guesswork that goes on too in this process. Then there is the element of luck as well. We said it can start as a suspicion or a conjecture and the attempt to prove it follows. When the proof succeeds then that statement gets elevated to the status of a theorem. All of these take time and above all the most important ingredient that is always there—hard work. Villani records this hardship in his book. When you read the theorem they formulated, you will see that it has many pre-

---

[2]Credits: ©Marie-Lan Nguyen/Wikimedia Commons.

Fig. 8.1 Cedric Vilani

conditions in order to be valid. They discovered these pre-conditions in the process of validating an idea they began. Without them, the idea would not work immediately by way of proof. However, when they constrained some conditions, it was at that time they saw the beauty of it and the proof worked. All of these experiences, good and bad, add to the elevation of a statement the status of a theorem.

Doing math is simply time consuming and hard work. There is no shortcut to this. The earlier you embrace the fact that your subject entails difficult work, the easier it actually gets.

The types of proof we have covered here are the ones you will need majority of the time. I hope you found the book helpful, instructive and makes you less afraid of mathematics. I hope this book has relieved you of mathematical phobia and relaxed you a bit more.

I give you my best wishes. I hope this little booklet has helped you in increasing your math skills, especially in that most dreaded area of proofs.

Any suggestions you may have to improve it will be gladly welcomed. Thank you.

References

1. H. DeLong, *A Profile of Mathematical Logic* (Dover Publications, New York, USA, 1998)
2. A.G. Hamilton, *Logic for Mathematicians* (Cambridge University, New York, USA, 1988)
3. A. Margaris, *First Order Mathematical Logic* (Dover Publications, New York, USA, 1990)
4. E.P. Wigner, *The Unreasonable Effectiveness of Mathematics in the Natural Sciences*. Communications in the Pure and Applied Sciences (1960), pp. 82–96
5. S. Meyer, *Mathematical Challenges to Darwin's Theory of Evolution*, https://youtu.be/noj4phMT9OE
6. C. Villani, *The Birth of A Theorem* (The Bodley Head, England, UK, 2011)

References

Index

© Springer Nature Switzerland AG 2021
L. P. Cruz, *Theoremus*,
https://doi.org/10.1007/978-3-030-68375-7

Printed in the United States
by Baker & Taylor Publisher Services